CIRIA C665

London, 2007

Assessing risks posed by hazardous ground gases to buildings

S Wilson
S Oliver
H Mallett
H Hutchings
G Card

CIRIA *sharing knowledge ■ building best practice*

Classic House, 174–180 Old Street, London EC1V 9BP
TELEPHONE 020 7549 3300   FAX 020 7253 0523
EMAIL enquiries@ciria.org
WEBSITE www.ciria.org

Assessing risks posed by hazardous ground gases to buildings

Wilson, S; Oliver, S; Mallett, H; Hutchings, H; Card, G

CIRIA

CIRIA C665   Updated © CIRIA 2007   RP711   ISBN: 978-0-86017-665-7

British Library Cataloguing in Publication Data

A catalogue record is available for this book from the British Library

Keywords		
Contaminated land, environmental good practice, ground investigation and characterisation, pollution prevention, urban regeneration		
Reader interest	**Classification**	
Ground, landfill and soil gas risk assessment, methane, carbon dioxide, hazardous ground gas, site investigation, detection, measurement and monitoring techniques, interpreting results from gas investigation	AVAILABILITY	Unrestricted
	CONTENT	Advice/guidance
	STATUS	Committee-guided
	USER	Land owners, developers (commercial and residential), professional advisors/ consultants (both engineering and environmental), builders and contractors, regulators (Environment Agency, Local Authority, building control) and other professional and non-specialist stakeholders

Published by CIRIA, Classic House, 174–180 Old Street, London EC1V 9BP, UK

This publication is designed to provide accurate and authoritative information on the subject matter covered. It is sold and/or distributed with the understanding that neither the authors nor the publisher is thereby engaged in rendering a specific legal or any other professional service. While every effort has been made to ensure the accuracy and completeness of the publication, no warranty or fitness is provided or implied, and the authors and publisher shall have neither liability nor responsibility to any person or entity with respect to any loss or damage arising from its use.

All rights reserved. No part of this publication may be reproduced or transmitted in any form or by any means, including photocopying and recording, without the written permission of the copyright-holder, application for which should be addressed to the publisher. Such written permission must also be obtained before any part of this publication is stored in a retrieval system of any nature.

If you would like to reproduce any of the figures, text or technical information from this or any other CIRIA publication for use in other documents or publications, please contact the Publishing Department for more details on copyright terms and charges at: email: **publishing@ciria.org** or **Tel: 020 7549 3300**.

Summary

Following the high profile gas explosions at Loscoe and Abbeystead during the 1980s, a number of reports were published in the 1990s on the measurement of ground gases, the assessment of risk such gases may present along with the measures that can be employed to mitigate these risks. More recent guidance has focused on licensed landfill sites but there is a need for up-to-date guidance relevant to existing or planned development.

With government policy encouraging redevelopment on brownfield sites, including those sites where there is a potential for the presence of elevated concentrations of potentially hazardous ground gases, the lack of clarity in the existing guidance has become more apparent. The resulting uncertainty is compounded by the lack of currency with respect to guidance on the methods of investigation, the adequacy of monitoring, the methods of risk assessment, and the selection of options for remediation. There is evidence that this current lack of clear and up-to-date guidance is leading to inappropriate, over- and under-conservative design, consequent resource inefficiencies and potentially unacceptable levels of risk.

This book gives up-to-date advice on all these aspects. The guidance it contains consolidates good practice in investigation, the collection of relevant data and monitoring programmes in a risk-based approach to gas contaminated land. A step-wise approach to risk assessment is described. Two semi-quantitative methods are set out for the assessment of risk:

1   For low rise housing with a ventilated under floor void at minimum 150 mm.
2   For all other development types.

Both methods utilise the concept of "traffic lights" to identify levels of risk. The mitigation and management of potentially unacceptable risk is described with reference to both passive and active systems of gas control (as well as brief reference to source removal etc).

Post development monitoring to confirm predicted behaviour is increasingly important in the remediation of contaminated land. However, in terms of development on or adjacent to gassing land, particularly for housing development, it is recognised that particular circumstances apply such that long-term/post construction monitoring would not normally extend beyond the point of sale or occupation of the development.

This guidance aims to ensure a consistent approach to decision making, particularly with respect to the need for, and scope of, remedial/protective design measures while remaining flexible enough to be relevant to site-specific and development variabilities.

Foreword

A series of guidance documents were published for the construction industry as part of CIRIA's 1990s research programme *Methane and associated hazards to construction*. These publications provided advice on the following:

- the occurrence and hazards of methane; CIRIA R130 *Methane: its occurrence and hazards in construction* (Hooker and Bannon, 1993)
- detection, measurement and monitoring techniques; CIRIA R131 *The measurement of methane and other gases from the ground* (Crowhurst and Manchester, 1993)
- interpreting measurements from gas investigations; CIRIA R151 *Interpreting measurements of gas in the ground* (Harries et al, 1995)
- protecting development from methane; CIRIA R149 *Protecting development from methane* (Card, 1996)
- strategies of investigating sites for methane CIRIA R150 *Methane investigation strategies* (Raybould et al, 1995)
- the assessment of risk CIRIA R152 *Risk assessment for methane and other gases from the ground* (O'Riordan and Milloy, 1995).

This book aims to clarify and update this earlier guidance, while being consistent with more high profile generic advice (eg the Environment Agency Model Procedures). The guidance aims to consolidate good practice in a risk-based approach to gas contaminated land and to ensure a consistent approach to decision making while remaining flexible enough to be relevant to site-specific and development variabilities.

Acknowledgements

The book was written by S Wilson (EPG Ltd), S Oliver, H Mallett, H Hutchings (Enviros Consulting Ltd) and G B Card (Card Geotechnics Ltd) under contract to CIRIA.

The authors

Steve Wilson, MSc BEng CEng CEnv CSci MICE MCIWEM FGS
Steve is the technical director of the Environmental Protection Group Limited and has extensive experience of risk assessment and remediation on contaminated sites, particularly on ground gas and vapours. He works as consultant to all sectors of the industry and often provides advice to local authorities on ground gas issues.

Sarah Oliver, BSc MRes
Sarah Oliver is a senior environmental consultant at Millard Consulting. She has substantial experience working on contaminated land projects with particular experience on landfill gas projects. Sarah acted as the main researcher and later project manager for the preparation of this guidance.

Hugh Mallett BSc MSc C.Geol SiLC
Hugh Mallett was a project director at Enviros Consulting during the preparation for the guidance and has since joined Buro Happold. Hugh has extensive experience many aspects of contaminated land and held overall responsibility for the delivery of the guide.

Heidi Hutchings BSc MSc AIEMA
Heidi is a consulting group manager within the Land and Water team at Enviros Consulting Ltd, involved in the project management, investigation and assessment of contaminated land prior to acquisition, reclamation or redevelopment. Heidi started her career 13 years ago as a landfill gas monitoring technician and since then has designed and implemented remedial works for a number of proposed and existing sites constructed on former landfills. Heidi also acted as project manager in the preparation of this guidance.

Geoff Card, PhD BSc CEng FICE EurIng CGeol CSi FGS
Geoff is the founding director of Card Geotechnics Limited. Geoff has some 35 years experience in geotechnical engineering and contaminated land development and works on major projects both in the UK and overseas. He is the author of CIRIA publication R149 *Protecting development from methane and associated gases* (CIRIA, 1996).

This project was funded by: English Partnerships, Roger Bullivant, Card Geotechnics Limited, Fairview Estates Ltd, RSK Group Plc, Taylor Woodrow, NHBC, A. Proctor Group, Waterman Group and CIRIA Core Members.

CIRIA, Enviros Consulting Ltd and the project team are grateful for help given to this project by the funders, by the members of the Steering Group, David Barry of DLB Environmental Ltd and by many individuals and organisations who were consulted.

Following CIRIA's usual practice, the research study was guided by a Steering Group which comprised:

Mr R Johnson (chairman)	Mouchel Parkman
Mr P Atchison	PA Geotechnical
Dr B Baker	Independent Consultant
Dr R Boyle	RSK Group Plc (now with English Partnerships)
Mr S Cassie	Worley Parsons Komex
Mr P Charles	CIRIA
Mr R Dixon	English Partnerships
Mr M Corban	A. Proctor Group
Mr C Douglas	A. Proctor Group
Mr R Douglas	SEPA
Dr B Gregory	Golders Associates
Mr R Hartless	BRE
Mr I Heasman	Taylor Wimpey Development Ltd
Mr J Keenlyside	Environment Agency (now with Jacobs UK Ltd)
Mr R Lotherington	Fairview Estates Ltd
Ms C MacLeod	Arcadis Geraghty & Miller International
Mr S Moreby	Gloucester City Council
Mr C Nelson	Zurich
Mr M Rafferty	NHBC
Mr N Rake	Roger Bullivant
Mr D Rudland	Halcrow
Mr R Shipman	ODPM
Mr P Witherington	RSK Group Plc

CIRIA's project manager for this guide was Miss J Kwan.

Contents

Summary iii
Foreword iv
Acknowledgements v
List of figures ix
List of tables x
List of boxes xi
Quick references xi
Examples xi
Case studies xi
Glossary xii
Abbreviations xvi

1 Introduction 1
    1.1 Background 1
    1.2 Legislative framework 1
    1.3 Scope and objectives 2
    1.4 Methodology 3
    1.5 Audience 3
    1.6 Structure of the publication 4

2 Hazardous gases 6
    2.1 Introduction 6
    2.2 Common sources of hazardous gases 8
    2.3 Physical and chemical hazards 10
    2.4 Nature of hazard 13
    2.5 Factors influencing the generation of ground gases and vapours ... 15
    2.6 Factors influencing the migration and behaviour of gases and vapours ... 16
    2.7 Ingress into buildings 19
    2.8 Ingress into other structures 22
    2.9 Summary 22
    2.10 Further information 23

3 Development of initial conceptual model and preliminary risk assessment ... 24
    3.1 Overview of risk assessment 24
    3.2 Desk study 26
    3.3 Initial assessment of risks 28
    3.4 Summary 30
    3.5 Further information 30

4 Methods of non-intrusive and intrusive investigation 31
    4.1 Setting the objectives of the ground investigation 31
    4.2 Investigation strategies and techniques 32

	4.3	Number and location of monitoring/sampling points 38
	4.4	Construction of monitoring/sampling points 40
	4.5	Summary 44

5 Monitoring methodologies 45
	5.1	Types of instrumentation appropriate for monitoring 45
	5.2	Selection process 50
	5.3	Monitoring methodologies 50
	5.4	Flux box measurements 57
	5.5	Number, frequency and duration of monitoring 59
	5.6	Presentation of data 61
	5.7	Summary 63

6 Sampling methodologies 64
	6.1	Sampling for laboratory analysis 64
	6.2	Methodology 66
	6.3	Analytical techniques to identify the source gas and hazardous properties 68
	6.4	Summary 70

7 Interpretation of results 71
	7.1	Background 71
	7.2	Understanding the collected data 72
	7.3	Refining the conceptual model 76
	7.4	Additional Phase II site investigation 77
	7.5	Summary 80

8 Assessment of risk 81
	8.1	Introduction 81
	8.2	Risk assessment process 83
	8.3	Methane and carbon dioxide 84
	8.4	Assessing vapours from hydrocarbon contamination 96
	8.5	Radon 97
	8.6	Summary 98

9 Remedial options 99
	9.1	Setting remedial objectives 99
	9.2	Philosophy for gas protection: basic concepts 100
	9.3	Passive and active systems of gas control (interruption of the migration pathway) 101
	9.4	Details of systems available 103
	9.5	Monitoring and alarms in buildings 107
	9.6	Summary 108

10 Post development monitoring 109
	10.1	Construction monitoring 109
	10.2	Existing guidance 109
	10.3	Future changes to the development and impact on soil gas regime ...110
	10.4	Risk perception issues 112

	10.5	Practical aspects of the long-term verification monitoring 113
	10.6	Recommendations for post construction/post remediation monitoring 113
	10.7	Verification/completion reporting 114
11	**Recommendations for research** **116**	
12	**References** **119**	
Appendices	. . . . **129**	
	A1	Trace components 129
	A2	Aquifer protection 130
	A3	Soil gas monitoring proforma 131
	A4	Available laboratory based gas analysis 133
	A5	Quantitative risk assessment 134
	A6	Modelling and risk assessment for vapours 158
	A7	Situation B derivations of ground gas emissions to calculate gas screening values (GSVs) for low-rise housing development with a ventilated under floor void (150 mm) 167

Bibliography **172**

List of figures

Figure 1.1	The process of managing risks related to hazardous ground gases . . . 5
Figure 2.1	Key gas ingress routes 20
Figure 2.2	General vapour intrusion schematic 22
Figure 3.1	Example of a schematic conceptual site model 28
Figure 4.1	Examples of shallow investigations 34
Figure 4.2	Examples of gas well response zones 43
Figure 5.1	Flux Box 59
Figure 6.1	The application of investigation methods to gas source identification . . . 69
Figure 7.1	Landfill gas generation against time 75
Figure 7.2	Estimation of landfill gas generation rates with GasSim 76
Figure 8.1	Risk assessment process for methane and carbon dioxide 84
Figure 8.2	Extent of gas monitoring against possible scope of protection 95
Figure 8.3	Complexity of risk assessment against remediation costs 96
Figure 8.4	Migration from a leaking underground tank 96
Figure 9.1	Available techniques for gas protection 100
Figure 9.2	Installation of HDPE membrane 101
Figure 9.3	Example of poor membrane installation – use of offcuts with insufficient sealing 102
Figure 9.4	Example of poor membrane installation – debris below membrane creating pressure points 102
Figure 9.5	Installation of positive pressurisation system and completed control panel 106
Figure A5.1	Developing a fault tree analysis 137

Figure A5.2	Example of simple fault tree for a gas cloud that could cause asphyxiation 137
Figure A5.3	Example of sensitivity analysis 140
Figure A5.4	Histogram of monitoring resulting 142
Figure A5.5	Landfill gas generation profile from GasSim Lite 144
Figure A5.6	Typical conceptual model 147
Figure A6.1	Temporal and spatial variability in vapour concentrations 161
Figure A6.2	Conception model based on the Johnson and Ettinger assumptions 166
Figure A7.1	Model low-rise residential property developed for calculating gas 168screening values for methane and carbon dioxide 168

List of tables

Table 2.1	Sources and origins of hazardous gases 7
Table 2.2	Physical and chemical properties of common hazardous soil gases .. 11
Table 2.3	Physical and chemical properties of some common volatile organic compounds involved in vapour intrusion 12
Table 2.4	General geology/soil gas migration links 19
Table 3.1	Summary of desk study information 27
Table 3.2	Initial generic conceptual model: Summary of environmental risks associated with hazardous gases for redevelopment to residential use 29
Table 4.1	Summary of exploratory techniques 35
Table 4.2	Spacing of gas monitoring wells for development sites 39
Table 4.3	Typical borehole spacings to detect off-site gas migration 40
Table 4.4	Examples of ground conditions and the suitable response zones ... 43
Table 4.5	Examples of significance of observations during investigations 44
Table 5.1	Gas monitoring instrument summary table 46
Table 5.2	Recommended data collection in monitoring programmes 52
Table 5.3	Summary of specialist monitoring techniques 55
Table 5.4	Summary of advantages and disadvantages of static/dynamic chambers 58
Table 5.5a	Typical/idealised periods of monitoring 60
Table 5.5b	Typical/idealised frequency of monitoring 60
Table 5.6	Idealised example: Summary of monitoring data 62
Table 6.1	Summary of pressurised sampling vessels 65
Table 6.2	Summary of non-pressurised sampling vessels 65
Table 7.1	An example of defined and up-to-date conceptual model summarising environmental risk associated with hazardous gases ... 79
Table 8.1	Classification of probability 85
Table 8.2	Classification of consequence 85
Table 8.3	Comparison of consequence against probability 86
Table 8.4	Description of risks and likely action required 86
Table 8.5	Modified Wilson and Card classification 88
Table 8.6	Typical scope of protective measures (risk management) 90
Table 8.7	NHBC Traffic light system for 150 mm void 92

Table 9.1	Development sensitivity	99
Table 9.2	Summary of passive systems	104
Table 9.3	Summary of active systems	106
Table 10.1	Summary of potential concerns	113
Table A5.1	Acceptability of risk	135
Table A5.2	Example of baseline and sensitivity parameters	141
Table A5.3	Factors affecting risk on gassing sites	156

List of boxes

Box 8.1	Situation A examples	88
Box 8.2	Owen and Paul risk assessment methodology	91
Box 8.3	Situation B examples	93
Box 8.4	Gas protection measures for low-rise housing development based upon allocated NHBC Traffic light	94
Box A5.1	Estimating membrane damage	152
Box A5.2	Ventilation design	153
Box A6.1	Examples of various methods in use for estimating vapour intrusion	158

Quick references

Potential hazards	25
Potential receptors	25
Potential pathways	26
A phased approach	31
Presence of made ground	40
Monitoring well considerations	41
Response zone considerations	41
Dual monitoring wells	44
Recirculation considerations	55
Field headspace test	56
Additional gas sampling considerations	68

Examples

Example 5.1	Example of typical gas monitoring round	53
Example 5.2	Internal gas survey with FID	57
Example 6.1	Use of a Gresham pump in gas sampling	67

Case studies

Case study 2.1	Offensive odours	14
Case study 2.2	Explosion incident caused by landfill gas with rapid drop of atmospheric pressure	17
Case study 2.3	Ingress of vapours	22
Case study 4.1	Significance of response zones	44
Case study 5.1	Example of typical gas monitoring round	61

Glossary

Aerobic	A process that involves oxygen.
Anaerobic	In the absence of oxygen.
Asphyxiant	A vapour or gas which causes unconsciousness or death by suffocation (lack of oxygen).
BOD	A measure of the potential for a polluting liquid to remove oxygen from the receiving water by biological or biochemical oxidation processes.
Borehole	A hole drilled in or outside the wastes in order to obtain samples. Also used as a means of venting or withdrawing gas.
Brownfield sites	A term generally used to describe previously developed land, which may or may not be contaminated.
Catalyst	A substance which speeds up a chemical reaction without itself undergoing any permanent change.
COD	A measure of the potential for a polluting liquid to remove oxygen from the receiving water by chemical oxidation processes (COD is always higher than BOD).
Combustion	A chemical process of oxidation that occurs at a rate fast enough to produce heat and usually light, in the form of either a glow or flames.
Cover	Material used to cover solid wastes deposited in landfills.
Concentration	The proportion of the total volume of void space occupied by a particular gas.
Conceptual site model	A representation of the characterisation of a site in diagrammatical and/or written form that shows the possible relationships between the contaminants, pathway and receptors. This shows the potential risks that the site poses given the intended operations and future use on the site.
Development	Works of construction, which may be buildings or civil engineering structures above or below ground, and including ancillary works, installations and open spaces associated with the structures.
Explosion	The bursting or rupture of an enclosure or a container due to the development of internal pressure from deflagration.
Factor of safety	This is used to provide a design margin over the theoretical design capacity to allow for uncertainty in the risk management process. The uncertainty could be any one of a number of the components of the risk management process, including calculations, material performance etc. The value of the safety factor is related to the lack of confidence in the process.
Flammable	A substance capable of supporting combustion in air.
Flux	Movement of gas.

Gas	One of three states of matter, characterised by very low density and viscosity (relative to liquids and solids), with complete molecular mobility and indefinite expansion to occupy with almost complete uniformity the whole of any container.
Gas chromatography	An analytical technique which can be used to identify and quantify substances according to the relative rate at which they separate out when passed through a specific medium
Gas flow rate	The volume of gas moving through a permeable medium or emanating from a standpipe per unit of time.
Gas generation	Volume of gas produced per unit mass or volume of waste per unit time.
Gas screening flow value	Gas concentrations (%) multiplied by the measured borehole rate (litres per hour).
Ground gases	A general term to include all gases (ie including VOCs or vapours) occurring and generated within the ground whether from made ground or natural deposits.
Hazard	A situation that could occur during the lifetime of a product, building or area and that has the potential for human injury, damage to property or the environment or economic loss.
Hazardous ground gases	Gas generated from ground which can cause adverse impact to human health, structures and the environment.
Hydrocarbon	A compound containing carbon and hydrogen only.
Ionisation	The process of changing a particle with no charge into one with a positive or negative charge, by the removal or addition of electrons.
Intrinsically safe	Of an instrument (or equipment) which does not generate an ignition source within the gas atmosphere being monitored (usually indicated by BSI accreditation).
Landfill	Waste or other materials deposited into or onto the land. After implementation of the Control of Pollution Act (1974), landfills became licensed and engineered.
Landfill gas	Gas generated from licensed, unlicensed, operating or non-operating landfill sites.
Leachate	The result of liquid seeping through a landfill and being contaminated by substances in the deposited waste.
LEL	The lower limit of flammability, that is the minimum percentage by volume of a mixture of gas in air which will propagate a flame in a confined space, at normal atmospheric temperature and pressure.
Limits of flammability	Concentration range bounded by LEL and UEL (see Abbreviations) within which a gas or vapour is flammable at normal atmospheric temperature and pressure.
Loscoe	A village in Derbyshire where a house was destroyed by a landfill gas explosion in 1986.
Made ground	Ground where there are deposits that have not been formed through natural geological processes. These may comprise combination of natural deposits together with products and materials, and waste produced by man.

Mercaptans	A group of organic compounds containing sulphur which have strong and unpleasant odours.
Methanogenic	Methane producing.
OEL	Limits related to personal exposure to substances hazardous to health in the air of the workplace.
Organic waste	Waste which contains a significant proportion of carbonaceous materials and is degradable waste (for example vegetable matter).
Oxidation	Reaction of species with an oxidant – normally, but not necessarily, "oxygen" from the air.
Paramagnetic	The property of substance which, when placed in a magnetic field, causes a greater concentration of the lines of the magnetic force within itself than in the surrounding magnetic field.
Passive control	The control of gas emission by provision of engineered venting pathways (for example wells and vent trenches) without mechanical aid.
Permeability	A measure of the ability of a medium to allow a fluid (gas or liquid) to pass through it.
Photo-ionisation	The ejection of an electron from an atom by quantum of electromagnetic energy.
Radius of influence	Area where changes happen.
Response zone	The perforated section of standpipe which allows gas in the unsaturated zone to enter a standpipe.
Recirculation	Method undertaken to measure the rate of gas recovery in a borehole.
Risk	The chance of a defined hazard occurring and achieving its potential.
Sampling	Collection of a portion of material for experimentation such that the material taken is representative of the whole.
Sensitive receptors	Property and structures which are more likely to be affected by a gas hazard either by virtue of their location or construction.
Soil gases	Gases present between soil particles or other sources which are generated from manmade and natural occurring activities. In this publication, soil gases have been used to describe all potential gases emission from the ground, including VOCs or vapours.
Solubility	The mass of the dissolved solid or gas which will saturate a unit volume of a solvent under stated conditions.
Source-pathway- receptor	This describes the linkage between the source of contamination and the pathway through which receptors can come into contact with contaminants.
Standpipe	A rigid tube inserted into the ground which allows the sampling of gas and water (usually in a borehole).
Surface emission rate	Volume of gas escaping from a unit area of ground in a unit of time.

UEL	The upper limit of flammability, that is the maximum percentage by volume of a mixture of gas in air, at normal atmospheric temperature and pressure, which will propagate flame in a confined space.
Vapour	Gases generated from volatile hydrocarbons.
Void space	The space between solid particles, occupied by a gas.
Zone of influence	The volume of ground surrounding a standpipe or other gas investigation installation which can or is being influenced by the use or control of that installation.

Abbreviations

ALARP	As low as reasonably practicable
AGS	Association of Geotechnical and Geoenvironmental Specialists
BOD	Biochemical oxygen demand
BRE	Building Research Establishment
BS	British Standard
BSI	British Standards Institute
CIRIA	Construction Industry Research and Information Association
CLEA	Contaminated land exposure assessment model
COD	Chemical oxygen demand
CQA	Construction quality assurance
Defra	Department for Environment, Food and Rural Affairs
DETR	Department of Environment, Transport and the Regions
DNAPL	Dense non-aqueous phase liquids
DoE	Department of the Environment
DPM	Damp proof membrane
DQRA	Detailed quantitative risk assessment
EA	Environment Agency
EAL	Environment assessment limit
EIC	Environmental Industry Commision
GSV	Gas screening value
HDPE	High density polyethylene
HPA	Health Protection Agency
HSE	Health and Safety Executive
LDPE	Low density polyethylene
LEL	Lower explosive limit
LNAPL	Light non-aqueous phase liquids
NAPL	non-aqueous phase liquids
NHBC	National House Building Council
NICOLE	Network for Industrially Contaminated Land in Europe
NRPB	National Radiological Protection Board
OEL	Occupational exposure limit
PPC	Pollution prevention control
ppm	Parts per million
SEPA	Scottish Environmental Protection Agency

SNIFFER	Scotland and Northern Ireland Forum for Environmental Research
UEL	Upper explosive limit
USEPA	United States Environmental Protection Agency
WEL	Workplace exposure limit
VOC	Volatile organic compounds

1 Introduction

1.1 BACKGROUND

Recent government policy has encouraged redevelopment (both residential and commercial) on brownfield sites including those where there is a potential presence of elevated concentrations of hazardous ground gases. Local authority duties under Part IIA of the Environmental Protection Act 1990 have also led to recognition of existing development which may be similarly located.

A number of reports were published in the early- to mid-1990s on the measurement of soil gases, the assessment of the risk such gases might present together with the measures that can be employed to mitigate such risks. More recent guidance has tended to focus on licensed landfill sites. Consequently, many construction professionals now engaged in the investigation and assessment of such sites have found a lack of clarity in the various guidance documents. This uncertainty is compounded by the lack of currency in the guidance with respect to methods of investigation, the adequacy of monitoring, risk assessment and remediation. Furthermore, there is evidence that the current lack of clear and up-to-date guidance is leading to inappropriate, over- and under-conservative design and consequent resource inefficiencies. This book provides up-to-date advice on all of these aspects and is relevant to both residential and commercial development, to existing properties and to the design of new build.

This publication and the advice it contains has been prepared to be generally consistent with CLR11 *Model Procedures for the management of land contamination* (Defra and Environment Agency, 2004a). The guidance presented here aims to consolidate good practice in the risk-based approach to gas contaminated land and to ensure a consistent approach to decision making while remaining flexible enough to be relevant to site-specific and development variabilities.

1.2 LEGISLATIVE FRAMEWORK

This publication provides technical guidance in an area where there is a complex mix of documentation related to legislative and regulatory procedures. It is not the purpose of this publication to describe in detail all regulations. However, it is important to recognise the context into which the guidance has to be employed.

Government policy is based upon a "suitable for use approach", which is relevant to both the current and proposed future use of land. When considering the current use of land, Part IIA of the Environment Protection Act 1990 provides the regulatory regime. The presence of hazardous ground gases could provide the "source" in a "pollutant linkage" which could lead the regulator (local authority or Environment Agency) to determine that considerable harm is being caused, or there is a significant possibility of such harm being caused to people, buildings or the environment. Under such circumstances, the regulator would determine the land to be "contaminated land" under the provisions of the Act, setting out the process of remediation as described in the DETR Circular 02/2000 *Statutory guidance on contaminated land* (DETR, 2000a).

In planning and development control of planned future use, Planning Policy Statement 23 *Planning and Pollution Control* (ODPM, 2004b) requires developers to undertake landfill gas risk assessments sufficient to demonstrate to the local planning authority that the proposals adequately mitigate any potential hazards associated with ground contamination, including gas. As described above, the Model Procedures provide a technical framework for applying a risk management process when dealing with land affected by contamination (including soil gas). The necessity of the risk assessment and risk management is also reiterated in a 2003 consultation document *Building development on or within 250 metres of a landfill site* (Environment Agency, 2003b).

Building control bodies such as Local Authority Building Control (LABC) and approved inspectors enforce compliance with the Building Regulations. Practical guidance with respect to the Building Regulations is provided by a series of Approved Documents, one of which, Approved Document Part C *Site preparation and resistance to contaminants and moisture* (ODPM, 2004a) aims to protect the health, safety and welfare of people in and around buildings, and includes requirements to protect buildings and surrounding areas from contamination including gas from the ground. The Approved Document includes a brief description of sources of soil gases, the risk assessment process and options for remediation.

Although this current CIRIA guidance is not designed to address the issues associated with gas derived from licensed landfills, recent publications related to landfills provide some useful and relevant information. In particular *Guidance on the management of landfill gas* (Environment Agency, 2004a) relates to the assessment of risks from landfill gases and provides a general overview of the process.

1.3 SCOPE AND OBJECTIVES

This publication is designed to provide up-to-date, practical and pragmatic advice on the assessment and mitigation of potentially hazardous ground gases with respect to buildings and their occupants. The advice covers both residential and commercial development and is focused on proposals for new build. However, it is also relevant to existing buildings and structures which may have been constructed without proper recognition of their foundation on a potentially hazardous gas regime. So the advice in this publication will be relevant to the regulator's assessment in carrying out duties under Part IIA of the Environmental Protection Act 1990.

The focus of this study has been on methane and carbon dioxide, although much of the text is also relevant to consideration of other contamination present in the vapour phase. However, it is not within the scope of this publication to address the specific issues of measurement, risk assessment and remediation of the wide variety of vapours that could be encountered in geo-environmental investigations. Where appropriate to both methane/carbon dioxide and to other contaminants in the vapour phase, generic comment has been made or advice given. However, it should be recognised that the presence of particular contaminants in the vapour phase will require specific consideration during investigation, risk assessment and in remediation. Such specific consideration is outside of the scope of this document.

The emphasis of this guide is on gases generated from the ground. Some of these are hazardous to human beings and the environment. Soil gas is a term that is used to specifically define gas present between soil particles.

The specific objectives of this guidance were to ensure that:

1   Valid data describing the gas regime are obtained:
    a   From appropriately designed and executed site investigation(s).
    b   From an appropriately designed and implemented monitoring programme.

2   The extent of uncertainty resulting from any limitations in the investigation and monitoring programme is properly recognised, and reflected in the risk assessment and design of remediation.

3   An appropriately rigorous, consistent and transparent strategy for the assessment of the risks posed by the soil gas regime is undertaken.

4   An appropriate strategy is developed for the mitigation of potentially unacceptable levels of risk in the:
    a   Assessment of the need for and scope of remediation.
    b   Design of any such remediation.
    c   Validation of approved remedial measures.

1.4 METHODOLOGY

The research undertaken to produce this publication comprised five tasks:

1   An initial scoping study.
2   The determination of good practice for the definition of the ground gas regime.
3   Identification of good practice in the risk assessment of ground gases.
4   Derivation of mitigation measures for the assessed level of risk.
5   Drawing this all together into practical guidance.

Existing guidance and information was examined by means of a focused and comprehensive literature search (see Bibliography). The state of current good practice was determined by means of a questionnaire survey sent to all members of the steering group, and members of both the AGS and the EIC. Consultation has been carried out with a total of some 100 organisations with expertise or experience of the assessment of risks associated with soil gases. This research took as its starting point and benefited from the results of the ground-breaking research in this area published mainly in the early- and mid-1990s. A number of the key references are identified in the bibliography.

During the period when this project was carried out (January to December 2005), research on related topics was also being carried out by the EIC (which will be published as BS 8485 *Code of Practice for the characterisation and remediation of ground gas in brownfield development*, in the near future) and by the NHBC (Boyle and Witherington, 2007). All the research contractors have cooperated fully to ensure that the results of these projects are complementary to, and generally consistent with each other, and that conflicting advice has not been produced.

1.5 AUDIENCE

This guidance has been written to be relevant to all parties/ stakeholders involved in the consideration of land and/or development (existing or planned) of land, affected by potentially hazardous ground gases and vapours. The target audience will include land owners, developers (commercial and residential), professional advisors/consultants (both engineering and environmental), builders and contractors, regulators (Environment Agency, local authority building control) and other professional and non-specialist stakeholders.

1.6 STRUCTURE OF THE PUBLICATION

Figure 1.1 illustrates the structured procedure for the assessment of hazardous ground gases. This structure is also reflected in the layout that has been adopted in this publication. The figure is a flow chart that has taken as its basis the framework from the Model Procedures (Defra and Environment Agency, 2004a). The publication follows this logical sequence; describing the tasks involved in the development of an initial conceptual site model and its relationship to hazards related to gases, then setting out the process and techniques necessary to clarify uncertainties or potentially unacceptable risks identified in this model. The methodologies available to assist in the assessment of risk associated with ground gases are then described. The assessed level of risk is then directly linked to the need for and scope of remediation, with the options for remediation set out in sequence from simple passive venting measures through to more complex multi-barrier and active systems. Guidance is also provided on the long-term monitoring and management of gas control systems.

The text is supported, where possible, by worked examples and case studies presented in boxes. "Quick reference" and "example" text boxes provide summarised advice. Detailed advice and information is also presented in a series of supporting appendices outside of the main text.

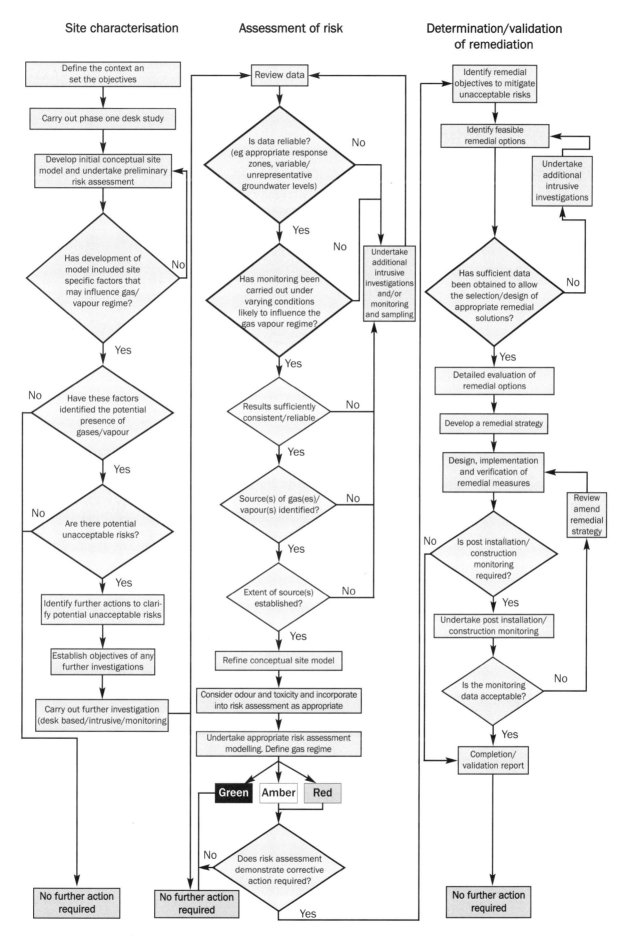

Figure 1.1 *The process of managing risks related to hazardous ground gases*

2 Hazardous ground gases

2.1 INTRODUCTION

Within the construction industry, the following hazardous ground gases are the most frequently encountered which require control and management:

- methane
- carbon dioxide
- radon
- hydrocarbon (including organic vapours).

Methane, carbon dioxide, radon and organic vapours occur naturally in the environment. In addition, buried organic matter has the potential to generate methane and carbon dioxide, as well as many other "trace" gases (see Appendix A1). There are numerous sources of these gases derived from anthropogenic and natural sources. Awareness of potential sources and options to differentiate between them is necessary to ensure appropriate long-term remedial solutions are applied to protect future development on the site. Table 2.1 summarises the potential sources of some hazardous gases, with typical concentrations of major constituents.

Methane is the principal hazardous gas for those sources involving the degradation of organic material. It is the most abundant organic compound in the Earth's atmosphere and is formed in many different environments (Hooker *et al*, 1993). Methane is biochemically reactive and is readily oxidised to carbon dioxide under aerobic conditions (in the presence of free oxygen and biochemical agents). However decomposition of organic compounds due to micro-organisms present in soil such as made ground, can also produce carbon dioxide. Carbon dioxide is often associated with the presence of methane, but can also be generated directly from soil.

Radon is a radioactive gas produced by radioactive decay of radium/uranium. It is present in all soils and rocks at variable concentrations. Hydrocarbon contamination is typically the result of spillage/leakage associated with industrial activity. Ground gas derived from organic degradation also often contains trace gases. The occurrence of these trace gases generally depends upon the nature of the organic material and degradation conditions, but may include substances with odorous, toxic, carcinogenic or other hazardous properties.

For the purpose of this publication the terms *soil gas* and *ground gas* are generally interchangeable and are used to describe all potential gas emissions from the ground whether from made ground or natural deposits.

Table 2.1 *Sources and origins of hazardous gases (Hooker et al, 1993)*

Source	Origin	Typical range of concentrations[2]		
		Methane	Carbon dioxide	Other (eg CO, H_2S, H_2)
Anthropogenic				
Landfill sites (include shallow and old landfill)	Microbial decay of organic materials derived from the disposal of putrescible materials	20 – 65%	15 – 40%	Several hundred trace organic gases (maybe odorous or toxic) (generally makes up <1% of total volume, eg H_2S
Made ground	Microbial decay of organic materials contained in reworked natural ground containing demolition and other wastes	0 – 20%	0 – 10%	
Foundry sands	Microbial decay of waste materials from the foundry process (phenolic binders, dextrin, coal dust, wood rags, paper)	Up to 50%	15 – 40%	Trace organic gases (generally <1% of total volume) (maybe odorous and/or toxic)
Sewage sludge, dung, cess pits/heaps	Microbial decay of organic materials	60 – 75%	18 – 40%	Trace organic gases (generally <1% of total volume) (maybe odorous and/or toxic)
Burial grounds (including cemeteries)	Microbial decay of organic materials contained within human/animal remains.	20 – 65%	15 – 40%	
Industrial/chemical/ petroleum sites/ manufacturing	Organic vapours derived from leaks or spills from storage, processing and disposal areas	30 – 100%[1]	2 – 8%	Trace organic gases (generally <1% of total volume) (maybe odorous and/or toxic), cyanide
Natural gas (supply pipes)	Leakage from bulk pipeline transportation of natural gas	90 – 95%	0 – 9.5% (carbon dioxide not present in mains gas however, methane may be converted following the leak)	1 – 27% C_2-C_4 alkanes, 4.7% CO
Natural				
Soil	Physical, chemical and biological transformations of rock during weathering	< 2 ppm	350 ppm	
Coal measures strata	Burial of vegetation under high temperatures and pressures, liberating gases as a by-product as a result of mining activities	<1 – 90%	0 – 6%	4 – 13% C_2-C_4 alkanes, 0 – 10% CO production of H_2S possible but rarely occurs in hazardous concentrations
Peat/bog areas	Gas formed by the microbial decay of accumulated plant debris under anaerobic conditions	10 – 90%	0 – 5%	
Alluvium (organic rich sediments)		0 – 5%	0 – 10%	
Radon-emitting rocks	Decay of naturally occurring uranium within soils and rocks	Variable	Variable	0-1000 Bq/m³ radon gas. Higher concentrations of gas up to 4 000 000 Bq/m³ have been recorded in the south west
Carbonate rich strata	Dissolution of calcium carbonate by acidic water	Variable	1 – 9%	
Other geological sources (eg oil and gas fields, oil shales, volcanic)	These are not especially relevant to UK but are relevant in some countries. (Further information can be found in CIRIA publication R130 *Methane: its occurrence and hazards in construction*).			

Notes:

Units: ppm – parts per million by volume (10 000 ppm = 1%)

1 Using bulk gas monitor reading as methane.

2 Typical range of concentrations quoted taken from published sources. The upper end of the range may be exceeded in particular sites/circumstances.

2.2 COMMON SOURCES OF HAZARDOUS GASES

There are a number of sources of ground gases, each of which is briefly described in Sections 2.2.1 to 2.2.11. Further details, including the typical compounds within each of these sources be found in Table 8.1 of *The monitoring of landfill gas* (Institute of Wastes Management, 1998).

2.2.1 Landfill sites

Landfill gas is a product of the biodegradation of organic materials contained in wastes deposited in landfill sites. Age and composition of landfill affect the gas regime. The gas regime will also be influenced by physical parameters such as volume/depth of waste and the groundwater regime, as well as environmental factors such as temperature, moisture content and pH value. These factors are considered in some detail in earlier CIRIA guidance (Barry *et al*, 2001). The Environment Agency *Guidance on the management of landfill gas* provides useful information on the mechanisms by which landfill gas is generated, its composition and physical and chemical characteristics and behaviour (Environment Agency, 2004a). Leachate from landfill sites may also contain dissolved gases or may degrade during migration to produce methane with carbon dioxide and associated gases.

Trace gases often associated with landfill gas include (among some 500 others) hydrogen, hydrogen sulphide and volatile organic compounds. These trace gases may derive directly from materials present in the waste but also from degradation processes. So the composition and concentration of trace gases vary with both the type and the age of the wastes. At sufficiently high exposure, these compounds can contribute to odour or health impacts. There has been much debate surrounding the potential health risks of living on or near landfill sites, and the measures that need to be taken to address any such risks. The Environment Agency has issued a list of priority compounds that should be measured in gases generated for landfills if a potential toxicity risk exists (see Appendix A1). Further information is contained in the Environment Agency's guidance document *Monitoring trace components in landfill gas* (Environment Agency, 2004b).

2.2.2 Made ground

Made ground will often contain degradable material such as wood, rags, paper and vegetation. However, the proportion of such carbon-rich materials is typically low, with major components often comprising re-worked clays, silts, sands and gravels together with anthropogenic inclusions such as ash, clinker, brick, concrete etc. Many brownfield sites contain made ground and on these sites the methane concentrations are usually not highly elevated, although there are exceptions, while concentrations of carbon dioxide can typically range to higher values. The rate of gas generation also tends to be low, resulting in small but sustained volumes of gas. There often tends to be a lack of driving force within made ground (see Section 2.6.1). The low rate of gas generation, the limited driving force and the fact that the gas is denser than air result in little upward migration of carbon dioxide.

It should be noted that some sites may contain a higher proportion of carbon- rich materials within the made ground. Elevated concentrations of methane related to the degradation of the materials can be detected on such sites. However, the rate of gas generation is still low resulting in small volumes.

2.2.3 Foundry sands

In foundry sands, organic materials resulting from the foundry process such as phenolic binders, detrin and coal dust, and other foundry wastes such as wood, lignin and paper can provide a substrate for methanogenic bacteria (Hooker *et al*, 1993)

2.2.4 Sewage sludge, dung pits/heaps

Methane and carbon dioxide are the main components associated with the anaerobic decomposition of organic components of sewage (Hooker *et al*, 1993). Hydrogen sulphide is also often present resulting from the degradation of organic matter and sulphur containing compounds (including mercaptans) in the sewage. Nitrogen oxide and ammonia gases are also associated with sewage. These gases can be a problem in sewer systems with confined spaces such as pipework, manholes and service chambers which can lead to potentially explosive, asphyxiating or chemically harmful atmospheres. Additionally the formation of sulphuric acid from the oxidation of hydrogen sulphide can corrode pipes, resulting in migration into the surrounding soils.

2.2.5 Burial grounds

The generation of gases from the decomposition of corpses is well documented (Polson *et al*, 1975). The gases generated are predominantly carbon dioxide and methane with trace amounts of odorous sulphur-containing gases. Diphosphane may be generated by anaerobic decomposition of phosphorus in skeletal material (generally in waterlogged areas). Other gaseous emissions may include formaldehyde, associated with the preparation of cadavers and present in medium density fibreboard (MDF), widely used to make coffins.

2.2.6 Spills, leaks, discharges

Petroleum hydrocarbons and other volatile organic compounds are associated with leakage and spillage. Hydrocarbons in the ground at high enough concentrations can pose a flammability or explosion hazard. Elevated concentrations are generally associated with strong odours and may be high enough to pose a health risk. However, it should be noted that very low concentrations of volatile organic compounds can also result in odours, which can be readily detected and be a driver to a risk assessment (see Section 2.4.3).

2.2.7 Natural gas supplies

Mains gas is derived from the same geological source as methane in coal mines. Leaks into surrounding soils can occur from damaged or poorly maintained underground pipes. In the UK, a combination of mercaphens and sulphide are added as odourants which can often be detected. Ethane additives will also indicate the presence of distributed main gases.

2.2.8 Mine workings

Methane is associated with coal bearing carboniferous strata, produced by the anaerobic decomposition of ancient vegetation trapped within the rock. Associated gases include higher alkanes (for example ethane), hydrogen and helium. Former shafts and/or fractured rock can provide a migration pathway to the surface and rising groundwater or flooding of mine workings can release trapped methane and carbon dioxide. Further details have been provided by the Department of Environment *Methane and other gases from disused coal mines: The planning response technical report* (Creedy *et al*, 1996).

2.2.9 Peat/bogs/organic-rich alluvial deposits etc

Methane from these sources is produced by the microbial decay of organic material under anaerobic conditions, usually waterlogged vegetation. Carbon dioxide is usually produced by acid reaction on carbonate fraction in any alluvial soil, and also generated by methane oxidation. Trace gases include hydrogen sulphide and light hydrocarbons. Methane can migrate large distances through soils. The source of the methane which caused the explosion at Abbeystead in 1985 was naturally occurring oil shales at more than 1 km depth.

2.2.10 Radon emitting rocks

Radon is a radioactive gas that occurs naturally and has no taste, smell or colour. It is formed from the decay of uranium, which is found in small quantities in all soil and rocks, in particular granite. Radionuclides (the decay products of radon) can damage lung tissues and ultimately lead to lung cancer. An action level of 200 Bq/m^3 was set by the former National Radiological Protection Board (NRPB), which recently merged with the Health Protection Agency to form the HPA's Radiation Protection Division <www.hpa.org.uk/radiation>.

Much information has been produced on radon, and the risks and mitigation measures are well defined. Further information on radon can be found in guidance published by the Building Research Establishment (1999) and the Health Protection Agency website <www.hpa.org.uk>.

2.2.11 Carbonate-rich strata

Acidic waters such as rainwater can react with calcium carbonate (eg chalk and limestones etc) to form carbon dioxide. Elevated concentrations of carbon dioxide (>five per cent) have been detected in confined spaces particularly those associated with groundwater abstraction infrastructure such as pump houses, located in chalk areas.

2.3 PHYSICAL AND CHEMICAL HAZARDS

Knowledge of the physical and chemical properties of the substances concerned is important when attempting to assess the potential hazards associated with ground gases and vapours. Boyle and Witherington (2007) set out the physical and chemical properties of the most commonly encountered soil gases are summarised in Table 2.2. Information on VOCs has been compiled in Table 2.3. Many physical and chemical parameters are fundamental but there are several chemical-specific parameters, such as partitioning coefficients, which are subject to some variation in quoted values. For example, the ranges of values quoted for the Henry's Law Constant and organic carbon absorption coefficient (koc) for Trichloroethene (TCE) are, respectively, 0.274 to 0.476 and 4.13 ml/g to 158.5 ml/g. So when modelling a complex pollutant linkage such as indoor inhalation of vapours, sensitivity analyses are essential.

Table 2.2 *Physical and chemical properties of some common hazardous soil gases (adapted from Boyle and Witherington, 2007)*

Properties	Methane	Carbon dioxide	Carbon monoxide	Hydrogen sulphide	Hydrogen
Chemical symbol	CH_4	CO_2	CO	H_2S	H_2
Density (g/l)	0.71	1.98	1.25	1.53	0.09
Melting point (°C)	-184	-78.5 (subliming point)	-205	-85	-259
Boiling point (°C)	-164		-191	-61	-252.87
Colour	Colourless	Colourless	Colourless	Colourless	Colourless
Odour	Odourless	Odourless (acid taste)	Odourless	Rotten eggs. Sense of smell disabled at high (toxic) concentrations	Odourless
Flammability	Explosive in air at 5–15 %. Range decreases if CO_2 present. >25 % CO_2 will render non-flammable	Non-combustible	Explosive in air at concentrations of 12.5–74.2 %	Flammable at concentrations of 4.5–45.5 % in air	Explosive in air at 4–74 %
Solubility in water (at 25°C)	25 mg/l	1450 mg/l pH dependent	21.4 mg/l	4100 mg/l	1.62 mg/l* at 21°C
Formation	Anaerobic degradation of organic material	Oxidation and combustion of organic materials, respiration	Incomplete combustion of organic material. Indicator of underground fires	Anaerobic decomposition of organic matter containing sulphur	Anaerobic degradation of organic material
Reactiveness	Fairly inert except with chlorine or bromine in direct sunlight	–	Low	Moderate – atmospheric half life of 1–30 hours	
Toxicity effects on humans	Low. But at high concentrations (>33 %) can result in asphyxiation due to displacement of oxygen	High. Headaches and shortness of breath at 3 %. Loss of consciousness at 10 – 11 %. Fatality at 22 %. OELs 1.5 % (short-term) and 0.5 % (long-term)	High. OELs 200 ppm (short-term) and 30 ppm (long-term). EAL 350 µg/m³ (long-term) 10 000 µg/m³ (short-term)	High. Asphyxiant at 400 – 500 ppm. OELs 10 ppm (short-term) and 5 ppm (long-term). EAL 140 µg/m³ (long-term) 150 µg/m³ (short-term)	Low. But at high concentrations (>30 %) can result in asphyxiation due to displacement of oxygen
Toxicity effects on vegetation	Displacement of oxygen	Cause toxic reactions in root systems. Displacement of oxygen			

Notes

Occupational eposure limits (OELs) are legal limits not necessarily safe limits (Health and Safety Executive, 2002)

Environmental Assessment Levels are *"benchmarks of environmental impact or harm… Benchmarks are available for both human health and ecological receptors"* (the EAL above are from Environment Agency 2003, Table D4 EAL for air for the protection of human health).

The long-term Environmental Assessment Limits (EALs) have been derived from Health and Safety Executive EH40 8 hour occupational exposure limits (now called workplace exposure limits) and converted to an annual mean (Environment Agency, 2003c). Short-term EAL have been derived from HSE EH40 15 minute OEL limits converted to an hourly mean. EALs are not available for methane, carbon dioxide and hydrogen.

Process contribution (PC) is the concentration of any emitted chemical impacting on a nearby receptor (eg human). This is added to the atmospheric background of that chemical to determine the Total Predicted Environmental Contribution (PEC). The EAL referred to in Tables 2.2 and 2.3 are PEC limits which incorporate the PC.

Process contribution (PC) values less than 10 per cent of the short-term EAL and one per cent of the long-term EAL are regarded as being environmentally insignificant. If the long-term predicted environmental concentration (PC + background concentration) exceeds 70 per cent of the long-term EAL, or the short-term PC exceeds 20 per cent of the difference between the long-term background concentration and the short-term EAL, the values are regarded as being of potentially significant environmental impact and detailed exposure risk assessment is necessary (Environment Agency, 2003c).

Table 2.3 *Physical and chemical properties of some common volatile organic compounds involved in vapour intrusion*

	Tetrachloro-ethene (PCE)	Trichloro-ethene (TCE)	Vinyl Chloride (VC)	Benzene	Toluene	Ethyl benzene	Xylene (o,p,m)
Chemical symbol	C_2Cl_4	C_2HCl_3	C_2H_3Cl	C_6H_6	C_7H_8	C_8H_{10}	C_8H_{10}
Molecular weight (g/mol)	165.83[a]	131.39[a]	62.5[a]	78.11[a]	92.15[a]	106[d]	106.16[a]
Normal boiling point, TB (°C)	121[a]	87[a]	-14[a]	80[a]	111[a]	136[d]	138 to 145[a]
Diffusion coefficient in air (m²/s)	7.20E-06[a]	7.90E-06[a]	10.6E-06[a]	8.80E-06[a]	8.70E-06[a]	7.5E-06[d]	7.00E-06 to 8.70E-06[a]
Henry's Law Constant, H' (Pa·m³ mol⁻¹)	2128 (20)[a]	1044 (25)[a]	1960 (17.5)[a]	442.5 (20)[a]	537 (20)[a]	633.67[d]	493.3-534 (20)[a]
Vapour pressure (Pa)	1900 (20)[a]	8600 (20)[a]	333000 (20)[a]	9970 (20)[a]	3000 (20)[a]	638 (10)[d]	660-860 (20)[a]
Water solubility (mg/l)	149[a]	1070 (20)[a]	1100 (20)[a]	1770[a]	535 (25)[a]	180[d]	160-180 (25)[a]
Diffusion coefficient in water (m²/s)	8.20E-10[a]	9.10E-10[a]	12.3E-10[a]	9.8E-10[a]	8.6E-10[a]	7.8E-10[d]	1.00-8.44E-10[a]
Critical temperature, TC (K)	620.1[a]	573.2[a]	432.01[a]	289[a]	591.8[a]	617[d]	617.0-630.3[a]
Log organic carbon partitioning coefficient, log KOC (ml/g)	2.53[a]	2.29[a]	0.6[a]	2.13[a]	2.25[a]	2.64 log cm³g⁻¹[d]	2.63-2.69[a]
Long-term workplace exposure limit WEL 8 hour ave mg/m³	345[b]	550[b]	3 parts per million[b]	1 part per million[b]	191[b]	441[b]	220[b]
Short-term WEL 15 min ref. period, mg/m³	689[b]	820[b]	Undefined	Undefined	574[b]	552[b]	441[b]
Long-term environmental assessment level µg/m³	3450[c]	1100[c]	159[c]	16.25[c]	1910[c]	4410[c]	4410[c]
Short term EAL µg/m³	8000[c]	1000[c]	1851[c]	208[c]	8000[c]	55200[c]	66200[c]
Principal hazard	Toxic to central nervous system, liver and kidneys, Group 2A (probable) carcino-gen[e]	Toxic to central nervous system, liver and kidneys, Group 2A (probable) carcino-gen[e]	Toxic to central nervous system and carcinogenic[e]	Toxic and carcinogenic[e]	Toxic[e]	Toxic to nervous system, liver, kidneys and eyes[e]	Very flammable, toxic to brain, lungs, causes eye, nose, throat and skin irritant[e]

Notes:

a   Values have been taken from the draft Environment Agency document *Review of the fate and transport of selected soil contaminants in the soil environment* (Earl *et al*, 2003). The temperature in °C is provided in brackets where applicable. These values are for illustrative purposes only and are not provided for use in a detailed quantitative risk assessment (DQRA) or similar assessment. In order to obtain appropriate values for DQRA or similar assessment it is recommended that a thorough literature search is undertaken and that the parameters are evaluated on an individual contaminant basis.

b   EH40/2005 Table 1 *List of approved workplace exposure limits*.

c   Environment Agency 2003 H1 *Environmental assessment and appraisal of best available technology* (Version 6). Table D4 EAL for air (for protection of human health).

d   Department of Environment, Food and Rural Affairs and the Environment Agency, *Soil guideline values for ethylbenzene contamination* (Science Report SGV 16 2004; updated 2005).

e   <www.en.wikipedia.org> then search for the compounds of interest.

2.4 NATURE OF HAZARD

The most commonly recognised hazards and effects of ground gases have been identified in CIRIA publication R130 as:

- flammable/explosive
- physiological effects
- odour
- effects on vegetation.

2.4.1 Flammable/explosive hazards

A summary of the potential flammable limits of hazardous ground gases is presented in Table 2.2. A hazard exists when soil gas accumulates which for methane in normal air is five per cent in a confined space at concentrations above the lower explosive limit (LEL). At standard atmosphere pressure and conditions this may be altered by other compounds in the ground gas (see below). Concentrations above flammable limits or upper explosive limits (UEL) can still be hazardous as dilution with air can easily reduce concentrations to within flammable limits.

For an explosion to occur, there should be:

1. A source of flammable gas.
2. A confined or enclosed space.
3. A source of ignition.
4. Sufficient oxygen to support combustion.

The flammability of gas mixtures is affected by the composition, temperature, pressure and nature of the surroundings. The flammability of methane will vary with changing concentrations of oxygen. If the oxygen concentration is reduced, the limits of flammability are reduced. For example, in air (oxygen 20.9 % v/v) the lower and upper explosive limits of methane are 5–15 % v/v, whereas at 13.45 % v/v of oxygen the lower and upper limits of methane are 6.5 and 7 % v/v respectively. At 13.25 % v/v oxygen, the mixture is incapable of propagating a flame (Hooker *et al*, 1993). In addition, the flammability of methane will alter with changing concentrations of carbon dioxide, with an overall effect of carbon dioxide altering the upper explosive limit of methane. More information is provided in Appendix C of the *Guidance on the management of landfill gas* (Environment Agency, 2004a).

2.4.2 Physiological effects

The physiological effects of different gases depend on the toxicity of that gas and the degree, nature and length of exposure. Trace components of gases derived from the degradation of organic materials can have toxic effects if present in high enough concentrations. The associated effects may appear in the short-term and/or in the long-term. Some physiological effects of exposure to most common ground gases are described in Tables 2.2 and 2.3. Occupational exposure limits (OELs) for both short-term and long-term exposure to many gases are provided by the Health and Safety Executive (HSE, 2002). Guidelines for vapour inhalation from exposure to contaminants in soil have also been published by the Environment Agency (Environment Agency *et al*, 2002a and Environment Agency *et al*, 2002b).

Radon, being radioactive, has the potential to damage cells and living tissue, with the potential to catalyse the on-set of cancer. The short-lived radioactive particles, created when radon decays, can remain suspended in the atmosphere. It is these particles that expose the lung to alpha radiation and increase the risk of developing lung cancer <www.hpa.org.uk>.

Migration of gas through the ground to a receptor such as inhabitants within a dwelling, without dilution, may cause toxicity thresholds to be exceeded. Many trace gases occur at such low concentrations that they will have little impact on human health, the environment or amenity. However, risks associated with exposure to hazardous chemicals should be specifically and individually assessed.

Any gas or mixture of gases will cause physiological effects if it displaces oxygen in a confined space, to an extent when oxygen concentrations fall below 18 % v/v. This begins with impairment of judgement (17 %), followed by anoxia and abnormal fatigue (10–16 %), nausea and unconsciousness (6–10 %) leading to death (< 6 %) (Card, 1996). Physiological effects can be more severe if carbon dioxide is the cause of oxygen depletion (see Table 2.2).

2.4.3 Odour

The major constituents of most soil gases are not themselves odorous (Table 2.2 and 2.3). However, the Environment Agency (Environment Agency, 2004b) identifies odorous trace components of gas generated from landfill include:

- hydrogen sulphide
- organosulphur compounds
- carboxylic acids
- aldehydes
- carbon disulphide.

Although there is no direct link between odour and adverse health effects, odours can cause symptoms such as nausea, and may increase a perception of adverse health effects. Furthermore, the hazard posed by a particular gas can also be at a lower concentration than the odour detection level. Detection of an odour (for example fuel or solvent) can often be a trigger for site investigation.

Case study 2.1 Offensive odours

Dwellings in a converted warehouse, East London

Naphthalene odour complaints to an environmental health officer (EHO) resulted in investigations of a bituminous damp-proof course and treated timber in the basement floor construction leading to Part IIA determination.

Hotel, north Wales

Odour of petrol vapours reported to an EHO triggered ground investigations that established a historical fuel spill on adjacent land leading to Part IIA determination.

Odour thresholds have been published for many individual gases and can be defined for specific gas mixtures via olfactometry. A major problem with odour assessment is that odour is subjective and individuals can be sensitised to particular odours below published thresholds. In addition, some gases are known to disable the sense of smell above certain concentrations, for example hydrogen sulphide.

2.4.4 Effects on vegetation

Strong correlations exist between high concentrations of methane and carbon dioxide, and vegetation die-back. This is principally due to the presence of gases causing oxygen deficiency in the root zone. Other hazardous gases which may affect vegetation include:

- methanol, formaldehyde and formic acid from the oxidation of methane
- trace components of gas generated for landfills (for example hydrogen sulphide, ammonia, benzene, ethylene, acetaldehyde and mercaptans).

2.5 FACTORS INFLUENCING THE GENERATION OF GROUND GASES AND VAPOURS

2.5.1 Source

Source-related factors include the size (mass) of the source materials, the proportion of biodegradable and other degradable material present, the age of the waste (and its degradable portion) and the presence of moisture/water. The conversion of a contaminant into the vapour phase (known as volatilisation), can occur directly from a free phase non-aqueous phase liquid (NAPL), from a dissolved phase plume or from a soil-adsorbed phase.

2.5.2 Timescale

When assessing the soil gas regime, it is important to determine whether the hazardous gas is being replenished by active generation or whether the gas is a residue which will deplete over time.

2.5.3 Chemical/biological factors

Many reports, including CIRIA publication R152 *Risk assessment for methane and other gases from the ground* (O'Riordan *et al*, 1995), set out parameters that influence the rate of decomposition of organic material (including hydrocarbons) by microbial activity and the subsequent gas (methane, carbon dioxide) production. In general, the conditions conducive to landfill gas generation are:

- moist, damp conditions which encourage greater rates of organic degradation and consequent gas generation
- water infiltration eg from rainfall
- conditions which are, or close to anaerobic (for methane) – generation of carbon dioxide occurs in the aerobic phase as well
- high proportion of biodegradable materials such as proteins, lipids, cellulose, carbohydrates, lignin and volatile fatty acids
- pH value between 6.5 and 8.5
- temperature between 25°C and 55°C
- high permeability – loosely compacted wastes/soils
- the ratio of biochemical oxygen demand (BOD) and chemical oxygen demand (COD) in leachate may be an indicator of the level of biodegration that is occurring. A BOD/COD ratio greater than 0.4 indicates that microbiological activities within the waste are still active and the population of bacteria is likely to increase, which will give rise to increased landfill gas production. If the BOD/COD ratio is less than 0.4, such as in leachate from old landfills, this indicates the microbal activity is

declining and the gas generation has peaked and is declining (Ehrig, 1996). However, this may be susceptible to a large degree of uncertainty and sampling variations, and as a result should only be used in support of direct measurement of the ground gas regime.

Under the right conditions, natural microbial action in soil can transform biodegradable compounds, converting hydrocarbons, for instance, ultimately into carbon dioxide and water (under aerobic conditions, that is in the presence of oxygen) or methane and water (under anaerobic conditions, that is in the absence of oxygen). So it is not uncommon that methane concentrations in the borehole headspace may reduce with time as the surrounding zone of methane in the soil is oxidised. There are exceptions to these optimal conditions, for example, the optimum conditions for the breakdown of vinyl chloride are aerobic whereas for TCE they are anaerobic.

Many gases and/or vapours in soils undergo reactions which can change their composition. When assessing risk, potential reaction mechanisms should be considered. The composition of gas vapours can be altered by:

- chemical makeup of source material
- adsorption of certain constituents onto soil particles
- micro-organisms within soils
- chemical activity within soils
- pH
- presence of oxygen
- presence of trace metals.

2.6 FACTORS INFLUENCING THE MIGRATION AND BEHAVIOUR OF GASES AND VAPOURS

2.6.1 Driving force

For a gas to migrate away from its source, there must be a driving force and an available pathway. For migration to be sustained, the gas should be continuously replaced by new generation. There are three principal factors influencing gas to migrate:

- pressure differential (generation of gas from within the source and changes in atmospheric pressure)
- diffusion along gas concentration gradients
- flow, in dissolved form, within liquids.

Further information on the movement of gases can be found in CIRIA publication R130 *Methane: its occurrence and hazards in construction* (Hooker *et al*, 1993) and CIRIA publication R149 *Protecting development from methane* (Card, 1996).

For a vapour, the main migration mechanism is diffusion, which is the movement of molecular constituents through a fluid in response to a chemical concentration gradient. That is, vapours will tend to redistribute in the wider soil from a region of high concentration to a region of low concentration.

The dominant influences controlling the migration of vapours are:

- density, dictating whether a liquid substance will float on groundwater (light non-aqueous phase liquid (LNAPL)) or sink to the bottom of an aquifer (dense non-aqueous phase liquid (DNAPL))

- viscosity, describing the resistance of the contaminant to flow through the soil (as a liquid or as a gas)
- solubility, as an indicator of a contaminant's potential mobility in water
- the vapour pressure
- as a result of the vapour pressure, the way in which the substances apportion themselves into different states (free product, dissolved liquid, or vapour) or environmental media (groundwater, soil, or soil gas).

Further information on the movement of vapours can be found in CIRIA publication R130 *Methane: Its occurrence and hazards in construction* (Hooker *et al*, 1993).

2.6.2 Meteorological conditions

Meteorological conditions can vary greatly over both short timescales and seasonally. Atmospheric pressure, rainfall, temperature and wind can all impact upon gas/vapour migration (Hartless, 2000). The influence of these factors is briefly discussed below.

2.6.3 Atmospheric pressure

A change in barometric pressure is a major influencing factor on gas emission and migration. At falling pressure increased emission rates occur as the gas increases in volume. Conversely, rising pressure causes air to flow into the ground, diluting soil gas concentrations. The rate of change in barometric pressure is also important. A swift drop over a small range has the potential to release a greater concentration of gas than a gradual drop over a greater pressure range which was the case in the Loscoe incident (see Case study 2.2).

Solubility of gas also increases with pressure, which could result in lower concentrations within the ground as more gas will be dissolved in water. Conversely, a pressure drop could cause the release of dissolved gas from groundwater into pore spaces and subsequently into atmosphere or along migration routes. Critical pressure changes are well described by Boltze and De Freitas (Boltze *et al*, 1996).

Case study 2.2 Explosion incident caused by landfill gas with rapid drop of atmospheric pressure

Loscoe, Derbyshire

At Loscoe, gas migrating from a recently capped landfill in coal measures caused an explosion which demolished a bungalow and severely injured the two occupants in 1986. The explosion was induced by the central heating boiler starting up in the early morning after a significant portion of the lowest part of the bungalow had accumulated an explosive mixture of landfill gas and air in the previous few hours.

From meteorological records it was clear that the few hours before the explosion had shown rapid decrease in atmospheric pressure in advance of a weather front crossing the site. Over a short period ie seven hours, there was a large drop of atmospheric pressure (which rapidly fell by 29 mb with hourly drop ranging between 3.3 mb and 4.8 mb). This illustrates the greater egress of gas during periods of rapidly falling pressure, particularly when atmospheric pressure has been relatively high and stable during the previous days and weeks.

Further details can be found in *Report of the non-statutory public inquiry into the gas explosion at Loscoe, Derbyshire, 24 March 1986* (King et al, 1988).

2.6.4 Rainfall and frozen ground

High rainfall will cause shallow groundwater levels to rise, resulting in a reduced available pore space in which soil gas can exist. Some of the gas will dissolve. However, the reduced pore space will result in an increase in gas concentrations, and an increase in potential release of the gases into the atmosphere.

Conversely, heavy rainfall will increase soil moisture content and can often cause temporary sealing of the ground surface, particularly in fine grained soil. This may cause a build up of gas and a greater potential for lateral migration. When the surface dries out, a faster release of gas may initially occur through cracks and fissures until a state closer to equilibrium is reached (Boyle and Witherington, 2007). Snow and ice can also seal the ground reducing the potential for gas to vent into the atmosphere.

2.6.5 Temperature

The solubility of a gas will usually increase with decreasing temperatures. However, some substances such as benzene, can behave differently. Changes in temperature will affect the density of a gas and increase the vapour release from liquid contaminants. In general, the higher the temperature, the more mobile the gas becomes.

Boyle and Witherington (2007) state that temperature changes (daily and seasonal) will also have an impact on the rate of biological activity, which is responsible for gas production. However, little work has been undertaken on the magnitude of this effect.

2.6.6 Wind

Pressure gradients can be formed between gas in the ground and that in the atmosphere by the effects of wind (the Venturi effect). High wind flows across a surface cause a pressure difference, resulting in the movement of gas from the soil to atmosphere. The direction of wind can also be a significant factor (for example if the ground is domed or undulating).

2.6.7 Ground conditions

Vegetation

Vegetation will have a slight impact on gas emission concentrations from the ground by altering the wind patterns close to the ground surface, reducing the significance of any Venturi effect, and thereby reducing the potential release of gas from the ground due to this process. Photosynthesis and respiration also impact the interaction of gases. However, these impacts are minor (Boyle and Witherington, 2007). Soil oxidation processes although also dependent on soil bacteria, can affect the interaction of gases (eg the oxidation of methane to produce carbon dioxide) see Hooker *et al* (1993).

2.6.8 Geology

The presence of soil gas and the movement through the ground is principally affected by the ground conditions. Permeable geology allows migration pathways for gas to flow (Table 2.4). Conversely, low permeability strata such as clays can inhibit and/or provide a barrier to gas migration. This is because high gas pressures may be required to move water held in soil pores and, indeed, any gas may remain trapped in the pore space. Strata will also affect the attenuation and subsequent absorption, degradation or decay of volatile liquids and consequential vapour release.

Table 2.4 *General geology/soil gas migration links*

Strata type	Examples	Migration pathways	Migration
Unconsolidated	Sand and gravel	Pore spaces	More generalised seepage of gas under low pressures for short distances
Consolidated	Chalk, limestone	Fractures, joints, bedding planes, fault lines	Gas potentially migrates under high pressure for longer distances within these isolated features

2.6.9 Anthropogenic influences

Human activities can create migration pathways (for example mine shafts, service runs, drains, building foundations, piles) and gas accumulation voids (for example inspection pits, basements).

2.6.10 Hydrogeology/tidal effects

To reiterate, the presence of groundwater will inhibit the movement of gases within the ground. Because vertical geostatic stress in a soil is higher than the lateral stress (due to self-weight of the overlying soil) water in the soil matrix will flow more easily in a horizontal direction than vertically. Gas bubbles will tend to form and combine in a horizontal plane as water in the soil matrix is displaced in the preferential horizontal direction. As more bubbles accumulate in the horizontal plane they link up forming networks resulting in open cracks or fissures, within the soil matrix, through which gas can flow with little resistance.

Some proportion of gas may also dissolve in the water. Rising groundwater will reduce the volume of gas within pore spaces resulting in increased gas pressure and release or lateral migration, known as the "piston effect". Changes in tide levels can result in changes in groundwater levels. The influence of tides on soil gases will depend on the ground conditions and distance to the coast or tidal river.

2.7 INGRESS INTO BUILDINGS

Soil gases can enter buildings via the following routes (as depicted in Figure 2.1):

- cracks or gaps in both solid and suspended floors
- joints formed during the construction process
- fractures in sub surface walls
- around service pipes and ducts
- wall cavities.

With the exception of specific joints, well constructed concrete slabs should not have cracks which can act as a pathway. However, there is always a potential for cracks to occur at any location across the slabs, generally as a result of induced stresses during or soon after construction, or from differential settlement or damage during use. Cracks can also occur at the floor/wall perimeter from the construction method, or as a result of shrinkage or building movement.

Typical traditional residential buildings use suspended floors over a void that was generally ventilated to avoid damp. A modern day recommendation – when building on gas contaminated land – is to construct suspended floors overlying ventilated voids (Building Research Establishment, 1991; Building Research Establishment, 1999; Johnson, 2001). However, whereas new construction often includes the use of plastic membranes to seal the floor and underlying void (Krylov *et al*, 1998), previous building methods incorporated the use of timber floors where the gaps between floor boards present a clear migration pathway for any gases in the underlying ground. Building Regulations now advocate the use of membranes in suspended pre-cast concrete floors to exclude hazardous gases and minimise potential pathways. In suspended timber floors, however, a membrane (together with a robust concrete oversite (over bare soil)) should be installed at ground level.

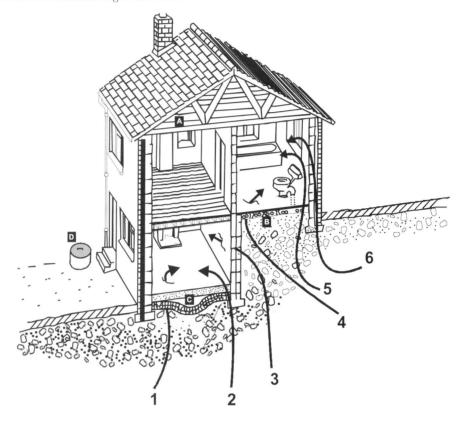

Key to ingress routes:

1   Cracks and openings in solid concrete ground slabs due to shrinkage/curing cracks.
2   Construction joints/openings at wall/foundation interface with ground slab.
3   Cracks in walls below ground level possibly due to shrinkage/curing cracks or movement from soil pressures.
4   Gaps and openings in suspended concrete or timber floors.
5   Gaps around service pipes/duct.
6   Cavity walls.

Locations for gas accumulations:

A   Roof voids.
B   Beneath suspended floors.
C   Within settlement voids.
D   Drains and soakaways.

Figure 2.1   *Key gas ingress routes (after Card, 1996 and BRE, 1991)*

As discussed above, there are many factors that can influence the migration of gases from within the ground to the surface. The development itself can also create pathways which alter the behaviour of soil gases. These may include:

- construction of piled foundations which may create a migration pathway linking a confined reservoir of gas, for example a peat layer, and the underside of the building. Similarly some ground improvement methods, for example the forming of vibro stone columns in the ground can also create highly permeable pathways for soil gas.

- surface pavings or other capping layers leading to accumulation of gas beneath the development and/or off-site migration
- pressure gradients created between the ground and building interior may encourage soil gases to migrate towards buildings. Such negative pressure relative to atmospheric can exist in a building as a result of:
    - **The Stack effect**: If the internal temperature in a building is higher than that outside, air is drawn into the building due to pressure differential, either through the external envelope of the building or through entry points in the ground floor construction. In a well-insulated building the air and soil gas is preferentially drawn in through the ground floor. In a heated building, warm air, including soil gas, rises through the stack effect which is then dissipated throughout the building. Further information can be found in CIRIA publication R149 *Protecting development from methane* (Card, 1996)
    - **The Venturi effect**: Positive air pressure occurs on the windward side of the building when exposed to wind pressure, whereas on the leeward side, suction occurs. So if there are openings on the leeward side, the internal pressure is reduced as air is drawn out. A pressure gradient develops between the inside and the outside. Soil gases may then be drawn into the building through entry points in the ground floor (see Figure 2.1).

The factors which influence the movement and mixing of gases in a confined space are:

- location of source relative to building
- existence of natural or artificial pathways
- gas density
- gas composition
- attenuation
- rate of ventilation with fresh air
- volume of confined space.

Further information on the ingress and behaviour of gases within buildings can be found in CIRIA publication R149 *Protecting development from methane* (Card, 1996). Further information on the ingress and behaviour of vapours can be found in draft guidance from USEPA (2002) (see Figure 2.2) and the guidance *Review of building parameters for the development of a soil vapour intrusion model* (Environment Agency, 2005a).

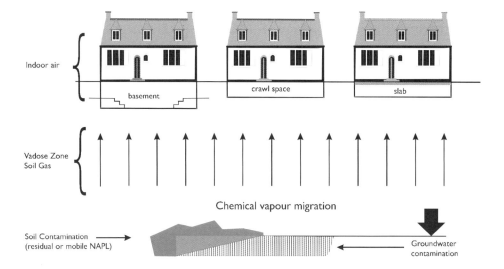

Figure 2.2 *General vapour intrusion schematic (USEPA, 2002)*

Once in a building, vapours can be attenuated (for example absorption in building materials and furniture) or diluted (via ventilation) before inhalation. Conversely, the presence of vapours can be obscured by the presence of elevated concentrations of the same vapours inside the building through the use of cleaning products, paint etc.

2.8 INGRESS INTO OTHER STRUCTURES

Ground gases and vapours can also accumulate within other structures present on a developed/development site. Construction workers are particularly at risk during building/drainage works where elevated carbon dioxide has accumulated. Particular areas of potential soil gas build-up are namely:

- piped drains and sewers
- soakaways/cess pits.

> **Case study 2.3   Ingress of vapours**
>
> **Redeveloped steelworks site, north Wales**
>
> The presence of a historical plume of volatile hydrocarbons in groundwater and residual soil contamination at its source (the former coal carbonisation plant) was established in a ground investigation that was prompted by the (explosive) discovery of BTEX vapours accumulating in storm-water drains.

2.9 SUMMARY

1 There are numerous sources of hazardous ground gases and vapours derived from anthropogenic and natural sources. Awareness of these potential sources and options to differentiate between them is necessary to ensure appropriate long-term remedial solutions are applied to protect future development.

2 Knowledge of the physical, chemical and biological properties of the substances of concern is important when attempting to assess the potential hazards, and migration potential associated with ground gas and vapours.

3 It is important to identify and understand the site-specific factors that may influence the behaviour and migration of gas/vapour.

2.10 FURTHER INFORMATION

Methane and other gases from disused coal mines: the planning response technical report, Wardell Armstrong (Creedy, D, Sceal, J and Sizer, K, 1996)

Interpreting measurements of gas in the ground, CIRIA R151 (Harries, C R, McEntee, J M and Witherington, P J, 1995)

Methane investigation strategies, CIRIA R150 (Raybould, J G, Rowan, S P and Barry, D L, 1995)

The measurement of methane and other gases from the ground, CIRIA R131 (Crowhurst, D and Manchester, S J, 1993)

Remedial engineering for closed landfill sites, CIRIA C557 (Barry, D L, Summersgill, I M, Greory, R G and Hellawell, E, 2001)

The disposal of the dead, 3rd Edition (Polson and Marshall, 1975)

Protecting development from methane, CIRIA R149 (Card, 1996)

Methane: its occurrence and hazards in construction, CIRIA R130 (Bannon, M P and Hooker, P J, 1993)

Guidance on the management of landfill gas (Environment Agency, 2004a)

Guidance for monitoring trace components in landfill gas (Environment Agency 2004b)

Radon: Guidance on protective measures for new dwellings, BRE Report 211, (Building Research Establishment, 1999b)

Health Protection Agency <www.hpa.org.uk>

Review of health and environmental effects of waste management. Phase 1 – Municipal solid waste and similar wastes (Defra, 2004)

Occupational exposure limits 1992 Edition, EH40 (HSE, 1992)

Guidance on the assessment and redevelopment of contaminated land Guidance Note 59/83 (ICRCL, 1987)

3 Development of initial conceptual model and preliminary risk assessment

3.1 OVERVIEW OF RISK ASSESSMENT

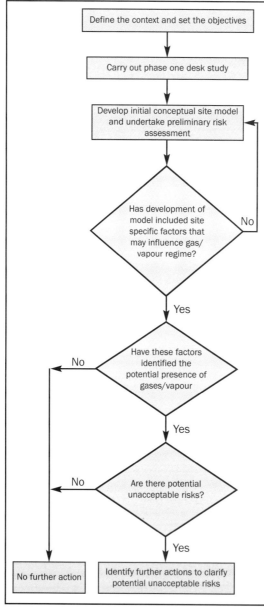

Note: Extracted from Figure 1.1

Risk assessment is the process of collating known information on a hazard or group of hazards to estimate actual or potential risks to receptors. For hazardous gases, the receptor may be humans, current or future buildings/structures or a sensitive local ecological habitat. Receptors can be connected with the hazard under consideration via one or several exposure pathways. Unless there is a pathway linking the source to the receptor, there is no risk. So, the mere presence of a hazard at a site does not mean that there will necessarily be attendant risks. This concept of a "pollutant linkage" (that is the linkage between a contaminant and a receptor by means of a pathway) is fundamental to the process of risk assessment in contaminated land, and also to the duties of local authorities under Part IIA of the Environmental Protection Act (1990) as described in the DETR Circular 02/2000 *Statutory Guidance* (DETR, 2000a).

As discussed in Section 1.6 the risk assessment process set out in this report follows the framework published in the Model Procedures (Defra and Environment Agency, 2004a) to which full and proper reference is recommended. The assessment of risk related to hazardous soil gases is discussed in Chapter 8 of this guide.

This chapter describes the factors to be considered in the development of the initial conceptual model for a site and what is termed in the Model Procedures as the "preliminary risk assessment".

3.1.1 Hazards

Potential sources of hazardous ground gases can be initially identified based on a review of the current and previous site uses undertaken as part of the desk study process. Not only the nature but also the likely extent of any hazardous gases needs to be considered, for example whether it is likely to be localised or widespread.

> **QUICK REFERENCE Potential hazards**
>
> Potential sources gases hazards include:
>
> - made ground
> - infilled pond
> - underlying natural strata (alluvial peat and chalk)
> - petrol re-fuelling area
> - off-site landfill
> - coal measures.

3.1.2 Receptors

The varying effects of a hazard on a particular receptor depend largely on its sensitivity. Receptors include any people including construction workers, animal or plant population, or buildings and structures within range of, and connected to, the source by the pathway. Receptors can, in addition, include construction materials that may be adversely affected by hazardous gases.

> **QUICK REFERENCE Potential receptors**
>
> Potential receptors include:
>
> - construction workers involved in redevelopment and further maintenance workers (in particular drainage workers)
> - current/future site occupiers and users
> - future buildings/structures
> - off-site occupiers
> - off-site buildings/structures
> - flora (both on- and off-site).

It is important to note that the nature of potential hazardous gases may be considerably altered due to activities carried out during construction and following development. For example, the development may entail capping the site with concrete hardstanding. This may cause gas to accumulate under pressure, or migrate via new pathways by creating a potential for off-site migration potentially affecting off-site receptors.

3.1.3 Pathways

The nature of the exposure pathway determines the "dose" which can be delivered to the receptor. The pathway which transports the hazardous gases to the receptor or target generally involves conveyance via soil, water or air. So it is important to identify the directness of the pathway, its length, the timescale for exposure and the potential for attenuation during migration.

> **QUICK REFERENCE Potential pathways**
>
> Potential pathways include:
>
> - migration via permeable strata
> - ingress into confined spaces
> - inhalation.

3.2 DESK STUDY

3.2.1 Introduction

The first and essential element of the risk assessment process is the desk study (Phase 1 risk assessment). A desk study involves the collation of information about the site to determine whether there is a potential risk from various sources including hazardous gases. Information gained at the desk study stage will enable the development of an initial conceptual site model and a preliminary risk assessment, and will assist in the efficient planning of subsequent intrusive investigations. The desk study should normally include obtaining, collating and assessing information about a site and its environs from a variety of sources (some publicly available and some which are likely to be private and/or confidential) for example Ordnance Survey maps. To provide further information supporting the desk study, it is important to undertake a walk over survey of the site. This will provide not only essential information on current site use, but also the likely nature of ground conditions (see Section 3.2.3). Further information on the scope of the desk study can be obtained from BS 10175 and BS 5930.

3.2.2 Objectives

At the outset of any land remediation project, the context of the contamination and the objectives of the remediation should be identified. Typical objectives of a desk study related to hazardous soil gas are to:

- assess the presence, extent and nature of ground gas source
- produce an initial conceptual site model and identify apparent principal pollutant linkages
- assess the implications of any identified environmental risks and liabilities associated with the site
- identify the likely ground conditions and identify potential locations for any subsequent intrusive investigations.

3.2.3 Site walkover survey

A typical site walkover survey should aim to:

- assess visual evidence of contamination (for example vegetation die-back)
- identify surrounding land uses and receptors in close proximity to the site
- for site in use, review site operation to assess the potential for pollution to occur such as:
    - interviewing appropriate staff
    - reviewing any environmental records held on-site.

Site visit questionnaires can provide a useful series of prompts during the walkover survey, as well as providing a permanent record of the observations of the visit. The questionnaires can be generic for certain types of site, for example operational industrial estates. However, at times there may be a need to develop site-specific questionnaires if particular aspects – such as process details – are required. Particular attention should be paid to those areas of the operation which potentially could present an environmental risk, and also those areas where compliance with environmental legislation is required. It is also critical to ensure that all relevant areas of the site are visited, and a plan with the actual boundary of the site is essential. Assuming permission is not withheld, a photographic record often also provides a useful part of the permanent record Table 3.1 summarises the scope of desk study information.

Table 3.1   *Summary of desk study information*

Information	Importance/relevance	Typical information sources
Site history	Identifies the history of site and surroundings. In particular identifies areas where: i. Potential hazardous gases may be present – infilled quarries/pits/ ponds/docks, landfill sites, coal mines/ shafts, reclaimed marsh areas, cemeteries, sewage works. ii. Potential hazardous vapours – tanks, works/factories, refineries etc may exist.	Ordnance Survey, local library, military records, planning records, aerial photos, local residents, Department of Environment industrial profiles (for contaminants related to historical use)
Geology and hydrogeology setting	Identifies the geology and hydrogeology of the site and surroundings. Identifies sensitive receptors (aquifers, rivers, ecology). Identifies potential for migration (permeable strata, joints, fissures, etc).	Geology and hydrogeology maps, Environment Agency records, borehole records, English Nature
Current land use	Identify potential infilled ground, vegetation dieback, current gas control/mitigation, spills/ leaks, above/below ground tanks, storage/ process areas, site receptors, surrounding land use/receptors.	Walkover survey, interviews with site personnel, site records, Department of Environment industrial profiles (for contaminants related to current use)
Pollution incidents	Location and nature of spills.	Local authority, Environment Agency
Waste disposal/ management licences and records	Type and location of wastes deposited on the site, material used to fill ponds, pits etc.	Local Authority (planning records), Environment Agency (licences), site records
Mining records	Location of mined areas, shafts/mine entries, spoil heaps.	The Coal Authority
Regulator consultation	Waste and gas management issues. Pollution events. Effectiveness of any remedial works.	Local authority, Environment Agency, English Nature, local petroleum officer
Underground services	Potential pathway and point of accumulation.	Utility companies, site records
Public rights of way	Identifies potential receptors that may be affected.	Ordnance Survey, local authority
Previous investigations/ monitoring/remedial works	Information to establish the ground conditions, soil gas regime and subsequent mitigation measures.	Local authority, Environment Agency, site records
Redevelopment plans	Identifies future receptors and implications to on-site and off-site receptors.	Developer/owner, local authority

3.2.3 Conceptual site model

Contaminated land risk assessment is based on development of a conceptual model for the site. This model is a simplified representation of the complex relationship between contaminant sources, pathways and receptors (known as pollutant linkages) developed on the basis of hazard identification.

An example of a simplified schematic conceptual site model is presented in Figure 3.1.

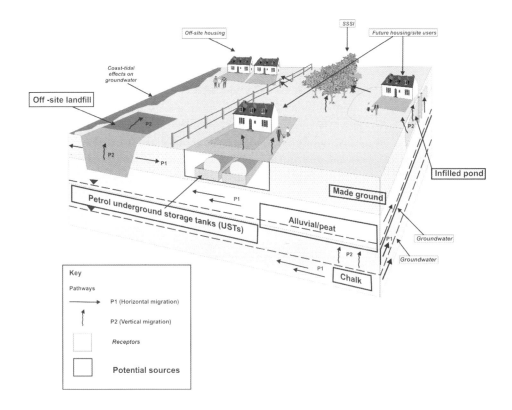

Note: This is a schematic diagram illustrating potential sources, pathways and receptors associated with a site. It is not intended to represent a detailed or precise site layout.

Figure 3.1 *Example of a schematic conceptual site model*

The development of a conceptual model is a critical first step to achieve a good understanding of current site conditions with respect to soil gas and in planning and scoping the subsequent tasks within the risk assessment process. A well-defined site-specific conceptual model will help identify data gaps and uncertainties and will be an essential tool in setting the data quality objectives for any ground investigation. It also facilitates the prediction of potential future risks – for example following development. The initial conceptual site model developed at the desk study stage should then be updated following subsequent site investigations, monitoring and sampling.

3.3 INITIAL ASSESSMENT OF RISKS

3.3.1 Exposure assessment

By considering the potential linkage between a source, pathway and receptor, an assessment can be made for each source of contamination on a receptor by receptor basis with reference to the significance and degree of the risk. The exposure risks can be assessed against the present site conditions during redevelopment or post development, or all scenarios.

Risk is based upon a consideration of both:

- the likelihood of an event (probability) (takes into account both the presence of the hazard and receptor, and the integrity of the pathway)
- the severity of the potential consequence (takes into account both the potential severity of the hazard and the sensitivity of the receptor).

The method of dealing with identified risks and the level of significance of those risks will be a function of site use. Table 3.2 considers some of the pollutant linkages represented in Figure 3.1 as an example of a preliminary qualitative assessment.

Note that health and safety during site investigation are discussed in Chapter 4.

Table 3.2   *Initial generic conceptual model: Summary of environmental risks associated with hazardous gases for redevelopment to residential use (please also refer to Table 7.1)*

Source	Pollutant	Receptors	Pathways to receptor	Associated hazard (severity)	Likelihood of occurrence	Risk
Off-site landfill	Methane, carbon dioxide	Future site users, construction workers	Migration, ingress and accumulation	Effect on human health (Severe)	**Low Likelihood:** Landfill completed in 1980, known to be unlined with no active gas extraction. Underlying strata may inhibit migration onto site. Potential for ingress of gas if buildings constructed along site boundary.	Moderate
		Buildings and structures	Migration, ingress and accumulation	Damage to building (Medium)	**Low Likelihood:** Landfill completed in 1980, known to be unlined with no active gas extraction. Underlying strata may inhibit migration onto site. Potential for ingress of gas if buildings constructed along site boundary.	Moderate/ Low
On-site made ground/ infilled pond	Methane, carbon dioxide	Future site users and construction workers	Ingress and accumulation	Effect on human health (Severe)	**Likely:** Pond filled during 1970s with waste derived from site. Vegetation dieback noted across area. Potential for exposure to construction workers during excavations. Potential for ingress of gas if buildings constructed over and around area.	High
		Buildings and structures	Ingress and accumulation	Damage to buildings (Medium)	**Likely:** Pond filled during 1970s with waste derived from site. Vegetation dieback noted across area. Potential for exposure to construction workers during excavations. Potential for ingress of gas if buildings constructed over and around area.	Moderate
		Off-site residents	Migration, ingress and accumulation	Effect on human health (Severe)	**Low Likelihood:** Pond filled during 1970s with waste derived from site. Vegetation dieback noted across area. Residential properties 20 m distant but underlying geology may inhibit migration.	Moderate
		Off-site buildings	Migration, ingress and accumulation	Damage to buildings (Medium)	**Low Likelihood:** Pond filled during 1970s with waste derived from site. Vegetation dieback noted across area. Residential properties 20 m distant but underlying geology may inhibit migration.	Moderate/ Low
On-site petrol refuelling area	Hydrocarbons/ volatile organic compounds (VOCs) and semi-volatile organic compounds (SVOCs)	Future site users and construction workers	Inhalation of vapours	Effect on human health (Medium)	**Low Likelihood:** Underground storage tank (UST) present for >30 years, no integrity testing, evidence of staining around delivery pumps. Potential for exposure to construction workers during excavations. Potential for ingress of vapours if buildings constructed over area.	Moderate/ Low
		Future site users and construction workers	Ingress and accumulation	Explosion (Severe)	**Low Likelihood:** UST present for >30 years, no integrity testing, evidence of staining around delivery pumps. Potential for exposure to construction workers during excavations. Potential for ingress of vapours if buildings constructed over area.	Moderate
		Future site buildings/ structures	Ingress and accumulation	Explosion (Medium)	**Low Likelihood:** UST present for >30 years, no integrity testing, evidence of staining around delivery pumps. No contamination observed during site visit. Potential for exposure to construction workers during excavations. Potential for ingress of vapours if buildings constructed over area.	Moderate/ Low
		Ecology	Direct contact, root uptake	Degradation of ecosystem (Medium)	**Unlikely Likelihood:** No contamination observed during site visit. No evidence of significant contamination having occurred. Hydrocarbon remediation known to have taken place. Present ecosystem well developed. No known impact from contamination.	Low
Natural strata alluvial/ peat deposits	Methane, carbon dioxide	Future site users and construction workers	Ingress and accumulation	Effect on human health (Severe)	**Low Likelihood:** Geology indicates potential presence of organic rich Alluvium. Gases within strata may migrate and ingress into buildings via foundations/services.	Moderate
		Buildings and structures	Ingress and accumulation	Damage to buildings (Medium)	**Low Likelihood:** Geology indicates potential presence of organic rich Alluvium. Gases within strata may migrate and ingress into buildings via foundations/services.	Moderate/ Low
Natural strata chalk	Carbon dioxide	Future site users and construction workers	Ingress and accumulation	Effect on human health (Medium)	**Unlikely:** Geology indicates presence of chalk at depth. Gases within strata may migrate and ingress into buildings via foundations/services. However, overlying groundwater may restrict upward migration.	Low

3.4 SUMMARY

The following tasks have been identified as the main stages of the preliminary risk assessment:

1. Define the context and objectives of the risk assessment.
2. Undertake collection and collation of desk study data and information.
3. Undertake a walkover survey (where permitted).
4. Outline the conceptual site model.
5. Identify potential unacceptable risks.
6. Identify pollutant linkage and further action to clarify potential unacceptable risks.

3.5 FURTHER INFORMATION

Environmental Protection Act 1990: Part IIA Contaminated Land, Circular 02/2000 (Department of the Environment, Transport and the Regions, 2000a)

Guidelines for environmental risk assessment and management (Department of the Environment, Transport and the Regions, 2000b)

Interpreting measurements of gas in the ground, CIRIA R151 (Harries, C R, McEntee, J M and Witherington, P J, 1995)

Methane investigation strategies, CIRIA R150 (Raybould, J G, Rowan, S P and Barry, D L, 1995)

The measurement of methane and other gases from the ground, CIRIA R131 (Crowhurst, D and Manchester, S J, 1993)

Contaminated land risk assessment – a guide to good practice, CIRIA C552 (Rudland, D J, Lancefield, R M and Mayell, P N, 2001)

Guidance on preliminary site inspection of contaminated land, Contaminated Land Research Report No 2 (Department of the Environment, 1994a)

Documentary research on industrial sites, Contaminated Land Research Report No 3 (Department of Environment, 1994b)

Code of practice for site investigations BS 5930:1999 (British Standards Institution, 1999).

Investigations of potentially contaminated sites – Code of practice BS 10175:2001 (British Standards Institution, 2001)

Industry Profiles (Department of Environment) <www.environment-agency.gov.uk/subjects/landquality/113813/1166435/?version=1&lang=e>

Model Procedures for the management of land contamination, Contaminated Land Report 11 (Defra and Environment Agency, 2004a)

Guidelines for combined geoenvironmental and geotechnical investigations (Association of Geotechnical and Geoenvironmental Specialists, 2000)

Potential contaminants for the assessment of land, R&D Publication CLR 8 (Defra and Environment Agency, 2002b)

4 Methods of non-intrusive and intrusive investigation

4.1 SETTING THE OBJECTIVES OF THE GROUND INVESTIGATION

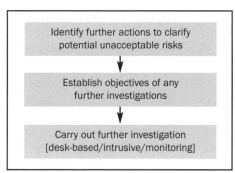

Note: Extracted from Figure 1.1

An overall objective for intrusive Phase II site investigations of potentially contaminated sites is:

"To provide information on actual and potential contamination and ground engineering characteristics to permit an assessment of environmental and physical risks and allow decisions to be made on the needs for and nature of any remedial work necessary for enabling a safe development." (Barry et al, 2001).

The site specific objectives of an investigation should be determined on the basis of:

- the conceptual site model developed by the Phase I desk study and walkover survey
- the implications of anticipated conditions for the use (or proposed future use) of the site
- the potential hazards and risks (including health and safety risk) which could arise from the site conditions.

The results of an investigation with appropriately defined objectives will enable further refinement of the initial conceptual site model and:

- establish the nature and extent of the current soil gas and/or vapour regime
- confirm the geological and hydrogeological conditions that may affect the soil gas regime
- assess the risk (qualitative and quantitative, if required, posed by the identified pollutant linkages
- identify the need for/scope of any further assessment/investigation
- establish the need for remediation works, the options for any such remediation and indicative costs.

> **QUICK REFERENCE A phased approach**
>
> Objectives should be kept under regular review and modified as appropriate to reflect the findings of investigations. A phased approach to the intrusive investigation enables the objectives to be reassessed as each phase is completed. Following the assessment, the objectives for the following phase can then be modified where required. A phased approach also ensures that the monitoring/sampling strategy of soil gases is flexible and can be adjusted in response to results or events on-site.

4.2 INVESTIGATION STRATEGIES AND TECHNIQUES

There is a wide variety of investigation techniques, which can be sub-divided into two general categories; intrusive and non-intrusive. Most Phase II site investigations involve the use of intrusive techniques, and it is generally not considered good practice to rely solely on non-intrusive techniques. However, a preliminary investigation (or even desk study) may consist of, or draw upon, non-intrusive techniques such as aerial photography or internal gas surveys of buildings, before designing an intrusive investigation. The results of non-intrusive investigation may often be utilised to aid in the design of targeted and cost-effective intrusive investigations.

There are many influencing factors to be considered when selecting an exploratory technique. The decision becomes more complex when the gas monitoring is part of a wider ground investigation. Consideration needs to be given to each factor in turn. The main aspects for the most commonly encountered situations are summarised in Section 4.2.1. These aspects are described in more detail in Sections 4.2.2 to 4.2.9 and Table 4.1 summarises the techniques available, and their advantages and limitations. The techniques have been grouped into "non-intrusive" and "intrusive".

4.2.1 Main considerations

The following text summarises the main considerations for the most commonly encountered situations where there is a need to carry out a risk assessment. This situation relates to gases from a landfill site or deposit of potentially methanogenic material where there is a sensitive receptor. A sensitive receptor may also be introduced to the situation through a planning proposal over or in the vicinity of the site. In such circumstances the following should be considered:

- the importance of the desk study
- the use of combined geotechnical/gas/water monitoring boreholes for the establishment of source of gas and process of generation
- the location and design of boreholes, eg selected to establish geological/geotechnical context of the site, nature of strata, likely permeability, nature, incidence, and lateral extent and depth of degradable/methanogenic materials
- potential use of geophysical survey techniques (eg to confirm the extent of tipping and pit etc)
- the nature and permeability of underlying and surrounding strata to enable possibilities of migration pathways to be assessed
- the possible need for (and advantages of) carrying out the investigation in stages (a phased investigation)
- health and safety issues (see Section 4.2.2).

In a typical scenario (eg former landfill) the borehole locations for the investigation would be planned as follows:

- a central location to enable measurement of the worst case situation with reference to gas generation
- peripheral locations (with the benefit of geophysical, photographic, and historical records information) to confirm extent of tipping
- boreholes immediately outside any tipped area to assess potential migration and pathways at different depths especially with reference to variable strata and sensitive receptors

- shallow boreholes situated close to housing, occupied buildings and other sensitive receptors, or the proposed location of planned development etc, to further assess migration.

Please note that a programme of internal monitoring for gas entering any existing buildings would be undertaken in situations where addition to previous results from monitoring suggest that there is a potential issue relating to possible emergency action or remediation.

The information gathered from such a systematic procedure is necessary to characterise fully the situation in such a way that a meaningful risk assessment can be undertaken. Such a process would also cover what is the most commonly encountered general situation with reference to landfill sites dealt with by local authority contaminated land officers in the course of their regulatory responsibilities in the context of both planning application assessments and potential determinations under Part IIA regime.

4.2.2 Health and safety

Any investigation involving the disturbance of ground in which hazardous ground gases are suspected should be carried out with appropriate safety precautions. Guidelines for the safe investigation by drilling of landfills and contaminated land (Site Investigation Steering Group, 1993) covers all of the potential hazards associated with drilling into gassing sites and areas where potential hydrocarbon vapours are present (Note: A revision of this document is currently in preparation). British Standards (BS 10175:2001 *Investigation of potentially contaminated sites – Code of Practice* and BS 5930:1999 *Code of Practice for site investigations*) (British Standards Institution, 1999 and 2001) should also be referred to during the design and execution of the investigation. Further health and safety guidance are also provided for specific circumstances by CIRIA (Steeds *et al*, 1996) and the Environment Agency (Environment Agency, 2004a).

4.2.3 Depth required

To target the source of the ground gas, exploratory techniques should be suitable to attain the depth at which the source is considered to be located. This factor consequently influences the type of exploratory technique chosen. CIRIA publication R150 *Methane investigation strategies* (Raybould *et al*, 1995) summarises the potential situations in which a shallow or deep investigation will be required.

4.2.4 Shallow gas investigation

A shallow gas investigation (to a maximum depth of approximately 5 m) is likely to be required where surface development exists or on sites where there is a potential gas risk. Figure 4.1 shows examples of when shallow investigation will be required.

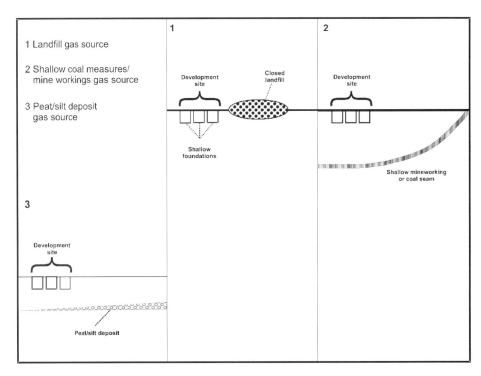

Figure 4.1 *Examples of shallow gas investigations (Raybould et al, 1995)*

4.2.5 Deep gas investigation

A deep gas investigation is likely to be necessary where:

- the gas source is at depth (for example coal measures or mine gas) or where the gas generating body is also deep (for example landfill)
- the proposed development involves piled foundations, or underground structures within potential gas-bearing strata at depth
- permeable or potential gas-transmitting strata are contiguous with a gas source and extend below the development site (for example mineworkings, or sand lenses which intercept waste or landfill)
- the proposed development involves the construction of deep buried structures (for example tunnels or pipelines) which pass through gas-bearing strata
- intervening geology and hydrogeology between a suspected deep gas source and the development site is not known (after Raybould *et al*, 1995).

As shown in Table 4.1, there are many different intrusive exploratory techniques which can attain varying depths. It is essential that the gas source and possible pathway(s) are taken into account in selecting the most suitable exploratory technique.

4.2.6 Flexibility

The exact source and location of the gas is not always known before a ground investigation is designed. For example, the depth of a landfill or made ground underlying a site may not be known until Phase I of the risk assessment is complete. In this situation, the flexibility of the investigation technique and the capability of the planned investigation to react to site-specific circumstances can be critical. There are many other factors which may require flexibility, for example the need to install deeper wells in areas of deeper made ground, or the need to excavate more locations in areas where further made ground is detected. As shown in Table 4.1, the flexibility of the exploratory techniques vary.

4.2.7 Sampling

Samples of soil gas should be collected only from appropriately installed gas monitoring wells (see Section 4.4).

4.2.8 Visual inspection required

It is important to determine and record the nature of the underlying ground conditions (for example visual and olfactory evidence) to assist in the definition of the soil gas regime. This will ensure an appropriate description of the soils/materials and help to identify potential sources of gas. As shown in Table 4.1, some techniques allow more visual assessment than others.

4.2.9 Timescale

The time it takes to complete a ground gas investigation will also reflect the investigation technique chosen. As shown in Table 4.1 the timescale of some exploratory techniques can range from more than 10 locations per day when trial pitting, to fewer than one deep borehole per day (eg cable percussion techniques). Although the timescale is an important consideration and can sometimes be an essential restriction of the project, it should be noted that the quicker and simpler intrusive techniques are not always suitably accurate and reliable.

4.2.10 Cost

The various exploratory techniques entail different costs (Table 4.1). Lower cost techniques may not provide the most reliable or accurate means of investigation. On large sites it may be possible to reduce the overall costs of an investigation and increase accuracy by the appropriate use of a non-intrusive or temporary technique before an intrusive permanent technique. For example, the use of either aerial false colour, infra-red photography, soil spiking or flux boxes may help to target areas of high shallow or surface gas concentrations.

Table 4.1 *Summary of exploratory techniques (after Raybould et al, 1995)*

Technique	Brief description	Advantages	Disadvantages	When to use
Non intrusive				
Aerial false colour infra-red photography	Based on different infra-red radiation absorption properties of healthy and stressed vegetation. Where methane present, the vegetation is often stressed. Not always apparent to naked eye but shows up when photographed using infra-red sensitive film. Note: film has not yet generally been superseded by digital techniques.	• able to assess wide area • possible to identify gas migration routes.	• requires air-borne cameras • not widely used • information not always accurate • vegetation may be distressed from causes other than methane such as leachate, incorrect use of pesticide, poor drainage • interpretation of results requires considerable skill.	Reliance solely on non-intrusive investigation techniques is not recommended. As part of a preliminary investigation/desk study, non intrusive investigation can aid in targeting the intrusive exploratory locations, especially on large sites. This may result in reducing the costs and time involved in the next phase. Preliminary/non-intrusive gas investigation should always be followed by an intrusive/confirmatory technique.
Aerial thermography	Infra-red scanner used instead of conventional camera to locate areas of underground combustion and methane generation. These areas will have a higher surface temperature.	• able to assess a wide area • possible to identify gas migration routes.	• requires air-borne camera • not widely used • information not always accurate • should be carried out at night to avoid interference from the sun • interpretation of results requires considerable skill.	Preliminary non-intrusive gas investigation should always be followed by an intrusive gas investigation.

Table 4.1 *Summary of exploratory techniques (after Raybould et al, 1995) (contd)*

Technique	Brief description	Advantages	Disadvantages	When to use
Non intrusive				
Satellite imagery/aerial photography	Based on aerial photography or locating satellite imagery for a "birds eye" view of the site in question.	• able to assess a wide area • enables confirmation of desk study information before the walkover survey • areas of distressed vegetation may be apparent.	• image alone cannot be used to assess gas migration routes/sources • aerial false colour infra-red images also required for interpretation of potential gas sources • not widely used.	A satellite image/aerial photograph alone would not be sufficient to assess potential gas sources/migration routes. However, such imagery can be useful to assess visually a large site before the site walkover. Preliminary non-intrusive gas investigation should always be followed by an intrusive gas investigation.
Near surface scanning over surface siource within FID	FID survey to ground (10 – 20 mm above ground) to detect surface emissions. Commonly used on landfill . Checking emission of landfill through the capping layer.	• no disturbance to ground, can identify emissions from fissures or other discrete pathways.	• survey does not identify the source of gas. The absence of gas concentration doesn't prove the absence of a soil gas source.	As part of desk study walk over or main site investigation. Should be followed up with boreholes or other form of monitoring.
Hyperspectral scanning	Based on the use of an optical scanning technique fitted to an aircraft.	• responds specifically to methane and carbon dioxide • can indicate the principal areas of gas emission and their scales.	• technique is not yet widely available • method produces only qualitative data and so further quantitative assessment is always required.	Hyperspectral scanning can be used to assess visually large sites. However, it can be used only to qualitatively assess the presence of methane and carbon dioxide. Preliminary non-intrusive gas investigation Should always be followed by an intrusive gas investigation.
Internal gas survey	Use of a portable flammable gas detector within buildings/ structures or in the area of services/voids.	• no disturbance to ground • possible to identify gas ingress routes • can identify emergency situation.	• survey does not identify the source of the gas • the absence of gas concentrations does not prove the absence of a soil gas source.	As part of a desk study site walkover (existing development). As a second phase of an intrusive investigation to examine the potential for gas ingress to existing development. Preliminary gas investigation should always be followed by an intrusive gas investigation.
Surface sampling (flux box)	An inverted container is placed on a prepared site surface and any gas collected is sampled through a valve over a recorded timescale.	• provides indication of gas emission rates • many locations/day.	• box can be disturbed • value of information is not certain • sealing of edges may be difficult • methodology for interpretation of data not standardised.	Useful as a second phase of intrusive investigation when gas concentrations and flows in the ground are high and further clarification is required on the potential for gas to be emitted.
Intrusive				
Soil spiking	Use of a metal spike to drive into the ground. The spike is removed creating a hole in which the soil gas can be monitored. Best practice for this technique includes placement of temporary flexible plastic probe with a gas tap into formed hole and temporary sealing of annulus at ground surface. Holes in the base of the probe allow monitoring from the bottom of the hole.	• rapid, quick and easy technique.	• a single "instant" reading most unlikely to provide reliable/ representative measure of soil gas regime • dilution of soil gas from ingress of air into the probe if surface not sealed after probe insertion • loose earth or water may clog the perforated holes of temporary pipe – more likely where underlying ground comprises clay or other low permeability material • driving spike can locally reduce permeability/inhibit ingress of soil gas into hole • cannot be relied upon for the measurement of gas flow or pressures • cannot visually assess the ground conditions • the probes are shallow and may not reflect deeper gas concentrations • loss of soil gas when spike is removed.	To provide evidence supporting presence of soil gas. Not be used to confirm or prove the absence of soil gas. Useful for identifying/delineating vapours for example fuel, solvent spills etc. as a pre-cursor to other intrusive investigation. Sole reliance on soil spiking data is not recommended. Further investigation with permanent installations would always be required.
Shallow/driven probes	Use of a hollow perforated pipe, driven into the ground sealed at the top with connection to gas detection device.	• permanent/semi permanent or temporary installation • quick and easy technique • can prove physically difficult to remove.	• maximum depth of 2 m (sometimes up to 4 m with use of sledge/trip hammer) • perforations can become blocked inhibiting gas ingress into pipe • can only be used to indicate gas is present • there is a potential for the arisings to be compacted affecting soil description.	To provide evidence supporting presence of soil gas. Normally not be used to confirm or "prove" the absence of soil gas. Useful for identifying/ delineating vapours for example fuel, solvent spills etc. as a pre-cursor to further intrusive investigation. Sole reliance on driven probes not recommended. Further investigation with permanent installations would always be required.

Table 4.1 *Summary of exploratory techniques (after Raybould et al, 1995) (contd)*

Technique	Brief description	Advantages	Disadvantages	When to use
Intrusive				
Hand auger	A hand-held auger with extendable sections is manually driven into the ground to form hole from which monitoring can be carried out.	• simple operation • allows deeper holes to be formed than soil spiking • allows visual assessment of the soil.	• A single "instant" reading most unlikely to provide reliable/representative measure of gas regime • can be physically difficult • can be time consuming • cannot penetrate difficult ground • potential dilution of gas during formation of hole.	Useful for identifying/ delineating vapours for example fuel, solvent spills etc, as a pre-cursor to further intrusive investigation. Sole reliance on hand auger results not recommended. Further investigation with permanent installations would always be required.
Driven probes (solid and hollow)	Use of hollow casing driven into the ground mechanically. Monitoring pipe is installed inside the casing, which is extracted leaving nose cone behind.	• limited ground disturbance compared to other permanent sampling • relatively quick (6 or more locations per day) • limited spoil produced • easily manoeuvred equipment, usually hand-held, preventing access difficulties • approximate maximum depth of 10 m.	• often limited penetration in strata comprising or containing gravels or cobbles (including broken rock such as chalk) or stiff clay • installation of gas wells can sometimes be limited due to hole collapse as or after the probe has been removed • compaction of surrounding soil may reduce gas migration into monitoring pipe.	Gas monitoring is possible but difficulties may arise in unstable stratum. Without the use of casing, installation of monitoring wells can sometimes prove difficult. Useful for identifying/delineating vapours, for example fuel, solvent spills etc, as a pre-cursor to further intrusive investigation.
Trial pits	Mechanical excavator used to create a trench/pit. A perforated standpipe is installed within pit, which is then backfilled with arisings.	• quick and easy technique with up to 10 or more trial pits excavated per day • generally deeper than hand-driven tools • also allows visual inspection of sample strata • can be used as part of overall ground investigation.	• maximum depth is limited by ground stability and reach of excavator • ingress of air and water from surface into pipe possible • causes ground disturbance which could cause difficulties on operating site • potential access difficulties for tracked excavator • may cause hazard to public health and danger to persons on site as it may bring contaminated material to the surface • disturbance of backfilled material may allow venting of gas • accuracy of gas source location reduced due to potential mixing of arisings during backfilling • hardstanding can slow down progress • backfilling can affect the local gas regime and can block or damage the installed pipe • soil disturbance can cause longer period before ground conditions stabilise • two wells often required to check readings due to considerable disturbance.	Not recommended. Data can be unreliable. May provide supporting evidence to presence of soil gas. Normally not used to confirm or prove absence of soil gases.
Boreholes (cable percussion, hand held window samples, track mounted window samples)	Use of a variety of techniques to form hole. Standpipe installed with gravel surround.	• use of gravel surround improves soil gas migration into borehole • greater depths achievable • minimal disturbance at ground level • can be used to monitor groundwater levels which may influence gas regime • allows visual assessment of arisings.	• may have access problems (dependent on rig type) • brings contaminated materials to the surface • may not be possible to install gravel surround on hand-held window sampler boreholes.	Installation of gas monitoring wells can be limited by hole collapse, as or after window sampler tubes are removed particularly in loose sediments or below perched water. The use of cable percussion allows deep gas monitoring wells to be installed.
Rotary boreholes	Hole is drilled by a rotary tool and flushed with air or water.	• normally quicker than cable percussion boreholes • greater depths achievable • relatively mobile rig • able to proceed through hard ground (eg rock) • may allow visual assessment of arisings.	• may have access problems • brings contaminated material to the surface • water flush can bring contaminated liquid effluent to the surface • air flush potentially hazardous where flammable gases and/or vapours are present due to sparks • arisings can be crushed by rotary movement of rig, limiting the visual assessment.	Caution should be taken when using air flush on sites with severe hydrocarbon impact, to ensure that vapours are not pushed into basements or other structures. Volatile analysis on soil samples may give unrepresentative results.
Use of trace gas	A unique trace gas is introduced into the monitoring wells at the suspected source. Monitoring of all wells can then be used to confirm that the suspected source is in fact the source and that the pathway exists.	• able to confirm the suspected source of gas • migratory pathways of some kilometres have been proven.	• the introduced gas will be detectable for a long time within wells, therefore the technique can be used only once • relies on the location of the monitoring wells being in the correct position.	Useful technique to confirm or refine the site conceptual model following the initial site investigation.

4.3 NUMBER AND LOCATION OF MONITORING/SAMPLING POINTS

The number and location of gas monitoring wells required for any site should be based on the site conceptual model and the need to provide appropriately robust data for assessment and design. The location and number of wells will therefore depend upon a number of site-specific factors. In principle, two basic approaches are available (Defra and Environment Agency, 2004a):

- targeted or judgmental sampling
- non-targeted or systematic sampling.

In practice, most gas sampling strategies will involve some combination of targeted and non-targeted sampling. Guidance on aspects of sampling strategy is provided in the CLR4 report (Department of Environment, 1994c) and CIRIA publication R150 (Raybould *et al*, 1995). It is recommended that, however small the site may be, a minimum of three wells will need to be installed. In addition, the following areas of a site will normally be targeted for gas well installation:

- critical areas of the site where the desk study has identified a higher risk of gas being present (for example historical infilled ponds or pits, the perimeter of a site nearest to a source of soil gas, a location between the gas source and a receptor, or within zones of permeable geology that could provide migration pathways)
- areas of developments (or proposed developments) that are sensitive to gas risk (for example below building footprints, service pathways)
- more commonly in areas of low risk or areas off-site to enable background concentrations to be collected and to confirm the absence of hazardous soil gas.

The spacing of gas monitoring wells is dependent on not only the location and number of potential gas sources, but also the sensitivity of proposed end use to soil gas ingress and the permeability of the ground (which will affect the radius of influence of the wells). For example, the spacing of wells will be different when monitoring on the gas source site from when assessing the risks posed by migration of the gas source.

Where the Phase I desk study has not identified particular areas of the site as priority or target areas (for example where the source is a homogeneous stratum below a site) then the most suitable method of setting out gas monitoring wells to give a representative indication of the gas regime is a uniform grid pattern (Raybould *et al*, 1995). The spacing of the wells should vary according to the specific site conditions and the magnitude of the risk associated with the gas source (see Table 4.2). However, it is important to recognise that the purpose of collecting soil gas data is to allow an assessment of risk and provide design data for gas protective measures. Therefore, as stated above, it is recommended that even for the smallest sites a minimum of three wells is installed.

Table 4.2   *Spacing of gas monitoring wells for development sites (Wilson et al, 2005)*

Gas hazard	Typical examples	Sensitivity of end use	Initial nominal spacing of gas monitoring wells[1, 2]
High	Domestic landfill sites	High[3]	Very close (<25 m)
		Moderate	Close (25–50 m)
		Low	Close (25–50 m)
Moderate	Older domestic landfills, disused shallow mine workings[4]	High	Close/very close (<25–50 m)
		Moderate	Close (25–50 m)
		Low	Close/wide (25–75 m)
Low	Made ground with limited degradable material, organic clays of limited thickness	High	Close (25–50 m)
		Moderate	Wide (50–75 m)
		Low	Wide/very wide (50–>75 m)

1   The initial spacing may need to be reduced if "initial" investigations suggest this is necessary to give a robust indication of the gas regime below a site. To prove the absence of gas closer spacing may also be required.

2   The spacing assumes relatively uniform ground conditions and the gas source present below a site. The spacing will need to be reduced if ground conditions are variable or if the investigation is an attempt to assess migration patterns from off site.

3   Placing high sensitivity end use on a high gas hazard site is not normally acceptable unless source is removed or treated to reduce gassing potential.

4   Petrol stations and other sources of vapours are most likely to be classified as gas hazard "Moderate", however site-specific assessment would be required.

A well network designed to investigate whether gas is present (detection monitoring) may be less extensive than one installed to determine the rate and extent of gas migration (assessment monitoring) or to monitor remedial activities. The following factors also need to be considered when using Table 4.2:

- for relatively consistent low generation potential sources, a wide spacing of wells may be sufficient (see Tables 5.5a and 5.5b)

- where zoning of gas protective measures is proposed on a site the number of wells required may have to be increased

- when assessing migration from an off-site source, close or very close spacing of wells may be required, particularly to demonstrate that gas migration is not occurring. Although strictly relevant only to licensed landfills, the Environment Agency's guidance document *Guidance on the management of landfill gas* may also provide some assistance for the design of investigations in other situations (Environment Agency, 2004a) (see Table 4.3).

Table 4.3 *Typical borehole spacings to detect off-site gas migration (Department of Environment, 1991 and Environment Agency, 2004a)*

Site description	Monitoring borehole spacing (m) Typical range
Uniform low permeability strata (for example clay); no development within 250 m.	50–150
Uniform low permeability strata (for example clay); development within 250 m.	20–50
Uniform low permeability strata (for example clay); development within 150 m.	10–50
Uniform matrix dominated permeable strata (for example porous sandstone); no development within 250 m.	20–50
Uniform matrix dominated permeable strata (for example porous sandstone); development within 250 m.	10–50
Uniform matrix dominated permeable strata (for example porous sandstone); development within 150 m.	10–20
Fissure or fracture flow dominated permeable strata (for example blocky sandstone or igneous rock); no development within 250 m.	20–50
Fissure or fracture flow dominated permeable strata (for example blocky sandstone or igneous rock); development within 250 m.	10–50
Fissure or fracture flow dominated permeable strata (for example blocky sandstone or igneous rock); development within 150 m.	5–20

> **QUICK REFERENCE Presence of made ground**
>
> It is recommended that whenever in deeper made ground (typically greater than 1.0 m) and organic matter or hydrocarbon spills are unexpectedly encountered during a ground investigation, additional soil gas monitoring wells (if not already planned) should be installed.

4.4 CONSTRUCTION OF MONITORING/SAMPLING POINTS

4.4.1 Monitoring well construction

Monitoring standpipes are commonly constructed of unplasticised polyvinylchloride (uPVC) or HDPE (high density polyethylene). HDPE is the preferred material as it is more resistant to attack by aggressive chemicals commonly encountered on contaminated sites. Other highly resistant materials, such as stainless steel, may also be used if severe contamination by aggressive organic contaminants (for example phenol, benzene) is known to be present at a site.

Typically monitoring standpipes are available in three diameters (19 mm, 25 mm and 50 mm). The selection of pipe diameter needs to take into account the method of borehole construction (for example the size of annulus and filter material, whether the gas of interest is methane/carbon dioxide or other potential contaminants in vapour phase and whether the installation is also to be used for groundwater sampling). Pipework is generally supplied in lengths of 1.0 m, 1.5 m or 3.0 m, and is either perforated or plain unperforated and screw-threaded to enable separate lengths to be joined together. Glue jointed pipework should be avoided as vapour from the adhesive can affect the measurement of the gas (both field instrument and chemical analysis). The standpipe is perforated at the time of manufacture (factory slotted) and slots are commonly less than 5 mm width.

> **QUICK REFERENCE Monitoring well recommendations**
>
> - subject to site-specific considerations, standpipes of 50 mm diameter should be used to allow consistency and interpretation at a later stage (see Chapter 8)
> - each length of pipe should be joined by screw thread
> - the use of adhesives or glue should not be permitted at any time due to the potential influence on the gas readings
> - in fine grained stratum, the slotted pipe should be taped into a geotextile sock to prevent the fine grains from entering the well
> - gas taps within a rubber bung, screw-on gas taps or push-on caps with gas taps should be used
> - each installation should be finished with a protective lockable cover (on operational sites it is recommended that all covers are flush to the ground to avoid damage to both the well and vehicles using the site).

4.4.2 Response zone

The response zone refers to the perforated length of standpipe which allows the gas in the unsaturated zone to enter the standpipe and collect in the upper unperforated length of the pipe. Most common installations comprise an unperforated length (usually not less than 0.5 m and no more than 1.0 m) near the ground surface and a perforated section below. Differing lengths of unperforated and perforated pipe can be used to dictate the soil horizon from which gas is monitored. The unperforated length near the surface should be fully sealed with bentonite clay surround to prevent surface water entering the response zone of the standpipe and to reduce the potential dilution of soil gases by atmospheric air.

Identified poor current practice consists of the use of one standardised response zone for each well regardless of the underlying site conditions or type of source present. Each response zone should be designed on a site-by-site or location-by-location basis. The box below provides a quick reference for the main aspects that should be considered when designing the response zone. Illustrations of example response zones can be found in CIRIA publication R131 (Crowhurst et al, 1993).

> **QUICK REFERENCE Response zone considerations**
>
> The depth of gas wells and selection of response zones should be based on characterisation of geological and hydrogeological conditions at the site, the presence of contamination and on the perceived level of risk associated with soil gas. The depth of wells should be sufficient to intercept any gassing sources or migration pathways that could affect the identified receptors.
>
> A common error is to conclude that there is no gas below a site when the wells have not been installed deep enough to intercept the source. For this reason, it is necessary to have experienced and appropriately qualified personnel supervising ground investigations/well installations to make decisions based on encountered ground conditions.

4.4.3 Equilibrium

When the borehole is initially constructed and the gas monitoring standpipe is sealed, the air composition within the standpipe will be in equilibrium with atmospheric air. So the initial concentration of gases such as methane and carbon dioxide will reflect the normal background concentrations in the atmosphere, typically less than 0.1 % v/v in air or lower. That is except where the well encounters a very active gas regime.

With time, however, soil gas from the soil pores will diffuse under concentration gradient or flow under pressure gradient into the standpipe. A high concentration of soil gas in the soil pores will mix with atmospheric air in the standpipe and will be diluted. Thus at any time the gas concentration is a function of the volume of the standpipe, the gas concentration gradient, the gas pressure gradient and the gas permeability of the soil. For soils of relatively high permeability, such as gravelly or sandy soil, equilibrium will be achieved relatively quickly. In clay soils of low permeability, the time period will be considerably longer.

4.4.4 Additional considerations

If there is more than one potential source of soil gas or discrete pathways a number of wells may need to have their response zones sealed into different strata rather than adopt the common method of installing a gravel surround along the whole length of the borehole. An example is given in Figure 4.2 where there are three potential sources of gas: a deep peat layer, a waste disposal (landfill) area and areas of made ground. Wells were installed and sealed into the different strata to determine the soil gas regime associated with each potential source. If the deep wells had not been installed then it is likely that the presence of gas below parts of the site would not have been identified.

In addition some wells were installed:

- above the peat layer to identify if any significant vertical migration
- outside of the landfilled area to identify any significant lateral migration.

Consideration should also be given to aquifer protection during the drilling and installation of monitoring wells on sites where known sensitive aquifers underly an impermeable layer, for example a non-aquifer clay layer or clayey made ground. A description of aquifer protection drilling techniques is included as Appendix A2. In addition, careful consideration should be given to the installation of a response zone to avoid creating a pathway.

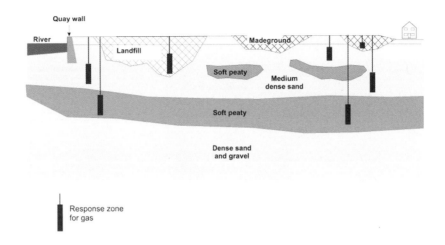

Figure 4.2 *Examples of gas well response zones (Wilson et al, 2005)*

Further examples of response zones have been summarised in Table 4.4.

Table 4.4 *Examples of ground conditions and the suitable response zones*

Target source	Ground conditions	Installation details
On-site landfill	0–0.5 m made ground	Plain pipe with bentonite surround (sealed/screened section)
	0.5–6.0 m landfill material	Slotted pipe with gravel surround (response zone)
	6.0–7.0 m clay	Bentonite surround
Deep made ground	0–0.5 m topsoil	Plain pipe with bentonite surround (sealed/screened section)
	0.5–3 m made ground	Slotted pipe with gravel surround (response zone)
	3.0–4.0 peat	Bentonite surround
	4.0–7.0 gravels	Bentonite surround
Peat	0–0.5 m topsoil	Plain pipe with bentonite surround (sealed/screened section)
	0.5–3 m made ground	Plain pipe with bentonite surround (sealed/screened section)
	3.0–4.0 peat	Slotted pipe with gravel surround (response zone)
	4.0–7.0 gravels	Slotted pipe with gravel surround (response zone)

Note: Response zone in bold

With location-by-location design of the response zone, it is essential to record all visual observations during the intrusive Phase II site investigation. Each observation could later be relied upon for justification of the response zone design. Table 4.5 shows examples of recorded observations and the significance of the information.

Table 4.5   *Examples of significance of observations during investigations*

Aspect	Examples	Significance
Stained/odorous contamination	Black staining of soils Oily staining of soils Oily sheen on groundwater Odorous soils/groundwater	Evidence of contamination and the strata in which it is located will aid in determining the horizon where vapour is more likely to be produced/stored which in turn aids in designing the response zone. The presence of contamination may also affect monitoring results.
Ground conditions/geology	Made ground (>1 m) Permeable gravels/impermeable clay Potential sources of soil gas (for example chalk, peaty layer)	The underlying geology can impact the gas migration pathways. It is prudent to know the geology in order to target potential sources with appropriate response zones.
Groundwater level	–	The response zone needs to be above the groundwater level in order to allow soil gas ingress.

Further information on the monitoring of odour and its significance is provided in *Guidance on the management of landfill gas*, (Environment Agency, 2004a) and *Horizontal odour guidance. Part 1: Regulation and permitting* (Environment Agency *et al*, 2002a) and *Part 2: Assessment and control* (Environment Agency *et al*, 2002b).

QUICK REFERENCE Dual monitoring wells

Separate wells are usually preferred, and the installation of combined gas and groundwater wells should always be considered carefully due to the potential for pathway formation and interference of the soil gas regime by soil gases dissolved in groundwater. Combined wells can be appropriate for example on sites where there is either no made ground or there is a requirement to monitor perched water within the made ground.

Case study 4.1   Significance of response zones

A ground investigation report was submitted to a local authority for approval for a site overlaying an historical sea inlet. Boreholes were installed, encountering thick peat at several metres depth, separated from the granular made ground by a clay layer with gravelly lenses. The response zones for the boreholes were in the shallow made ground. The developer argued that migration of gas to the proposed development was prevented by the impermeable clay layer. However the regulator determined that the response zone design had not taken in to consideration the recommended foundation construction of piles extending into the peat, or the gravel lenses offering potential migration routes. The development was delayed while additional wells with appropriate response zones were installed and monitored. Monitoring encountered relatively high concentrations of methane, and gas protective measures were installed.

4.5   SUMMARY

1   The definition of appropriate objectives relevant to the ground gas regime is critical to the investigation strategy and should be kept under regular review.

2   There are various non-intrusive and intrusive investigation techniques available. It is not recommended to rely solely upon non-intrusive techniques and some preliminary intrusive techniques.

3   The final design and location of gas monitoring installations should be based on actual ground conditions encountered during the intrusive investigation to ensure that all potential sources of gas are targeted.

4   It may be necessary to undertake a second phase of investigation to fully determine the nature and extent of any identified gas.

5 Monitoring methodologies

5.1 TYPES OF INSTRUMENTATION APPROPRIATE FOR MONITORING

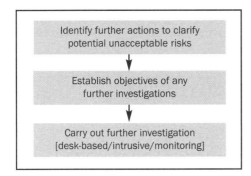

Note: Extracted from Figure 1.1

5.1.1 Overview

There is a wide range of ground gas monitoring instruments available, each of which has its own advantages and limitations. The range of instrumentation typically comprises the following basic elements:

- sensors to detect the gas (giving an analogue or digital signal)
- gas pump
- processing unit
- display for the operator
- data logger
- power supply.

It is important that the operator understands the sensor operation so that the correct instrument can be utilised for the monitoring and that all limitations and potential interferences are understood.

The most commonly used gas monitoring instrument is an infra-red analyser. The instrument is based on a limited number of sensors which are used both singly and in combination to enable methane, carbon dioxide, oxygen and other gases (typically hydrogen sulphide and carbon monoxide) to be recorded simultaneously. Commonly available sensors have been summarised in Table 5.1.

As an essential input to risk assessment, gas flow should also be a routine field measurement to accompany measurements of concentrations. Chapter 7 provides further information including an explanation of the different types of measurement that can be made.

Table 5.1 also summarises the available instruments for the measurement of gas flow.

Table 5.1 *Gas monitoring instrument summary table (Crowhurst et al, 1993)*

Equipment	Gas	Advantages	Disadvantages	How it works and when to use
Field gas concentrations equipment (or similar)				
Infra-red analysers	Methane, carbon dioxide	• detects a wide range of gases • wide and sensitive detection range • widely used and proven within field • gas sample passes unchanged through the sensor • additional instrumentation can be added to the instrument to measure trace gases and flow.	• readings affected by water in the system • selection of correct wavelength difficult • instruments may not be intrinsically safe for internal use • cross-sensitivity with hydrocarbons (watch out for high methane reading) • poor resolution at low concentrations of methane.	**Methodology** The gas sample being measured is stored in a cell within the analyser. Radiation is released and passed through the cell. The amount of radiation absorbed by the gas sample relates to the concentration of gas being measured (Crowhurst, 1987). **Application** The infra-red analysers are generally the preferred soil gas monitoring instrument for all sites where methane and carbon dioxide are suspected. The infra-red analyser should not be used for headspace analysis and internal gas surveys.
Flame ionisation detectors (FID)/Gastec	Flammable soil gases and/or vapours, hydrocarbon vapours	• widely used for detection of hydrocarbon vapours and flammable gases • highly sensitive (0–10 000 ppm) ideal for use as a primary investigating device for screening soils for volatile vapours • presence of elevated concentrations indicates need for further investigation • the FID can be used in open air • some versions of the FID can be used in confined spaces (intrinsically safe versions only).	• require a minimum amount of oxygen in the sample to operate correctly • some versions may not be intrinsically safe for confined spaces or in flammable environments • responds to any combustible gas or vapours • sample destroyed by sensor • can be affected by extreme temperatures and rain • transportation (for example by courier) difficult due to presence of hydrogen cylinder.	**Methodology** The FID measures the increase in ions produced when organic matter within the sampled gas passes through a hydrogen flame (Crowhurst, 1987). **Application** Soil vapour surveys (for example identification, delineation of hydrocarbon spills) Internal gas surveys of buildings and services to detect the ingress of methane It is important that the sensitivity of the FID being used is suitable for the gases that need to be measured.
Photo ionisation detector (PID)	VOCs, gasoline vapours, jet fuel, petroleum gas, benzene, other gases (most organic gases)	• detects VOCs at low toxic thresholds • intrinsically safe • newer models available with over 250 gases and gas mixtures available for detection • highly sensitive can detect 1 part per billion total non-methane VOCs • new model can be set to read over 250 gases.	• sample temporarily ionised by sensor • can be affected by extreme temperatures and rain • can be affected by cross contamination • reading affected by humidity: dessicant filter needed on wet/humid days • does not speciate: reads total non-methane VOC • reads accurately when one known gas is being sampled, to which the read out screen has been set. Data are semi-quantitative when reading complex gas mixtures • readings are affected by humidity: dessicant filter needed on wet/humid days.	**Methodology** An ultraviolet light lamp within the detector is used as a source of energy to remove an electron from neutrally charged VOC molecules. This produces an electrical current which is proportional to the concentration of gas/vapours/contaminant. **Application** For the indication of solvents in particular aromatic and chlorinated solvents. It is important to confirm that the lamp being used in the PID has sufficient energy to ionise the compounds of interest. The PID cannot detect methane.
Catalytic instruments No longer available	Flammable soil gases and/or vapours	• formerly widely used now largely superseded by infra-red instruments • detects presence of flammable soil gases and/or vapours accurately at low concentrations • responds to any flammable gases • relatively inexpensive.	• the sensing element may deteriorate with age, catalyst may be poisoned by minor constituents • the instrument may produce an ambiguous signal if the concentration of flammable gas greatly exceeds the lower explosive limit (LEL), indicating an apparently safe condition when it may not be safe • operation of instrument requires the presence of oxygen for oxidation to occur (12 to 15% v/v) • cannot distinguish between gases.	**Methodology** This sensor detects the presence of gas from the heat generated by the oxidation of the combustible material on a small heated sensing element (typically comprising platinum resistance thermometers and a catalyst). The signal produced is proportional to the concentration of flammable gas (Crowhurst, 1987). **Application** When it is known that only one gas is present (therefore little potential for cross contamination). When gas concentrations are relatively low/below the LEL.

Table 5.1   *Gas monitoring instrument summary table (contd)*

Equipment	Gas	Advantages	Disadvantages	How it works and when to use
Thermal conductivity	Flammable soil gases and/or vapours, carbon dioxide	• can detect concentrations up to 100% v/v • detector is easily combined with other detectors • fast response measurements.	• can be inaccurate if two or more gases present have similar conductivities • poor resolution at low concentrations.	**Methodology** Thermal conductivity sensors operate by measuring differences in thermal conductivity of the gas sample relative to air (or other reference gas). This is achieved by measuring the voltage change in a circuit generated by the change in resistance. The change in resistance is caused by the change in rate of heat loss as the sample passes over a heated sensing element (Crowhurst et al, 1993). **Application** For the measurement of flammable gases/vapours and methane only
Electrochemical cell	Oxygen, hydrogen sulphide, carbon monoxide, hydrogen	• can be used to detect a wide range of gases other than landfill gas • relatively inexpensive.	• cell has a limited life span • can be poisoned by certain gases • can lose sensitivity due to moisture, corrosion and poisoning • potential for interference of carbon monoxide by hydrogen.	**Methodology** An electrolyte is used to create a chemical reaction between the gas sample and electrolyte which in turn produces an electrical current. The sensor measures the current produced which increases as the concentration of gas increases. **Application** For the measurement of oxygen, hydrogen sulphide, carbon monoxide and hydrogen only.
Paramagnetic sensors	Oxygen	• no interference from majority of other gases • accurate measurements • robust in the field.	• accuracy can be affected by atmospheric changes • relatively expensive • prone to cross contamination.	**Methodology** Paramagnetic sensors measure the changes in the magnetic susceptibility in the gas sample. Oxygen has a positive magnetic susceptibility whereas all other gases have a negative magnetic susceptibility which enables the oxygen in a gas mixture to be measured (Crowhurst, 1987). **Application** For measurement of oxygen only, for example for health and safety monitoring in confined spaces.
Chemical detector tubes	Total hydrocarbons, carbon dioxide, hydrogen sulphide, carbon monoxide, gasoline/ petrol, diesel/jet fuel, nitrogen oxides	• used for a wide range of gases/ vapours • wide detection range • simple to use.	• time consuming if more than one gas being monitored • potential for low accuracy • prone to interference effects • only suitable for indication.	**Methodology** Chemical detector tubes contain one or a mixture of chemicals, which are reactive to certain gases. Gas sample drawn into tube and reacts with the chemical causing a change in colour. Either the resultant colour or length of colour in the tube can be related to a certain concentration of gas (Crowhurst et al, 1993). **Application** When there is a potential for toxic compounds. Indicator of soil gases and/ or vapours/toxic compounds. All readings should be confirmed using laboratory analysis.
Portable gas chromatography and portable mass spectometer (GC-MS)	Bulk gas/trace components	• GC-MS is available in a reduced size • now available in portable carry case • the instrument is robust for use in field • offers the accuracy usually only obtained within a laboratory • individual compound quantified.	• cost effective only for long-term monitoring • limited battery life • need access to mains or generator power.	**Methodology** Sample is injected into a gas stream, which results in each compound separating. Each compound leaves the stream at a specific time. As it leaves the stream, it passes through a detector. The response of the detector is related to the concentration of the compound. By comparing the response of the detector to known standards, the concentration of the soil gas in the sample can be calculated. The mass spectrometer is a type of detector for a gas chromatograph. **Application** This relatively new equipment offers the accuracy of the laboratory version but also mobility, thus when transported to site, the GC-MS can be ready to analyse samples within 30 minutes. However, to set up such equipment is expensive and would be cost effective only for long-term monitoring.

Table 5.1 *Gas monitoring instrument summary table (contd)*

Equipment	Gas	Advantages	Disadvantages	How it works and when to use
Semi-conductors	Flammable soil gases and/or vapours, hydrogen sulphide, carbon monoxide	• not easily poisoned • can be made very sensitive • long-term stability of measurements • good selection of some toxic gases.	• lack of selectivity to combustible gases • not specific to any one material • accuracy and response depend upon humidity.	**Methodology** This sensor measures the change in electrical resistance of a solid state semi-conductor as a gas sample passes through. The change in resistance is related to the concentration of combustible gas in a sample (Crowhurst, 1987). **Application** Generally unsuitable for environmental monitoring because it can react to any gas. Typically used for domestic purposes and leak detection.
Mass gas flow monitor	Direct measurement of flow rate	• measured directly at monitoring point • simple method • robust, hand held monitor • flow measurement not affected by wind, steam or contaminants.	• prone to dust and condensation build up • correction factors must be applied if the gas composition is significantly different from that of air.	**Methodology** This sensor directly measures the flow rate from a standpipe. **Application** This instrument is now widely used and accepted for the measurement of gas flow. The monitor can also be used to measure gas pressure in the borehole.
Gas flow pod	Direct measurement of flow rate using a gas analyser	• the flow pods comprise robust aluminium • the flow range is 0.1 to 12 l/h with a flow resolution of 0.1 l/h.	• the flow pod is not certified as intrinsically safe and cannot be used in confined spaces or in flammable environments.	**Methodology** This pod is available as an additional attachment to certain infra-red analysers. The flow pod directly measures the flow rate from a standpipe. **Application** This instrument is now widely used and accepted for the measurement of gas flow. Used in conjunction with gas analyser. Cannot be used in confined spaces or in flammable environments.
Rotameters, vane anemometers, bubble-flow meters	Direct measurement of volume flow rate	• measured directly at monitoring point • simple method • flow measurement not affected by wind, steam or contaminants.	• can be used only where gas quantity and pressure are high • calibration is affected by the density, temperature and pressure of gas mixture • can underestimate • prone to damage • correction factors must be applied if the gas composition is significantly different from that of air.	**Methodology** Rotameters consist of a graduated vertical tube with a taper towards the lower end. Flow is represented by the position of a float within the tube. Bubble-flow meters are based on the time taken for a soap bubble to rise up a graduated tube with a known volume. The volume is recorded and converted to the volume flow of gas. Vane anemometers operate by counting the number of revolutions per unit-time of a vane which rotates in the presence of a gas flow. **Application** These instruments have been widely used. All of them are generally more appropriate for use on a gassing landfill sites where the pressure and flow is greater. Soap bubble meters typically suffer from less resistance and are therefore more appropriate for sites with low pressure and flow. However, measurements can still be unreliable due to fluctuations.
Hot-wire anemometer	Gas velocity from monitoring point	• allows measurement of gas flow velocity.	• changes in composition of gas stream can give spurious results • easily influenced by cross wind and hand vibration.	**Methodology** This instrument is based on the principle of change in thermal conductivity caused by a gas flowing over a heated wire (typically platinum). As velocity of gas flow increases, the thermal conductivity of the gas increases and the hot wire is cooled. This produces a change in resistance which is then used to calculate the gas flow velocity. **Application** Hot-wire anemometers are commercially available. Not recommended for measuring landfill gas flows in the field because measurement is recorded above an open borehole and not through a gas tap. This measurement will therefore not be accurate enough for subsequent risk assessment.

Table 5.1 *Gas monitoring instrument summary table (contd)*

Equipment	Gas	Advantages	Disadvantages	How it works and when to use
Field gas flow equipment (or similar)				
Micromanometer	Atmospheric/ borehole pressure	• measured directly at monitoring point • relatively simple technique.	• cannot be related directly to an emission rate • measures gas pressure only within the borehole.	**Methodology** This instrument typically comprises a device with a transducer and two sample lines. One sample line measures atmospheric pressure while the other measures the pressure at the borehole. **Application** For the measurement of gas pressure within the borehole. This instrument cannot measure gas flow.
Autothermal desorption tubes (also known as passive diffusion tubes) (see EA LFTGN 04 *Guidance for monitoring trace components in landfill gas* (EA, 2004)	Speciated volatile organic compounds	• preserves sample intact and inexpensive • allows full speciation of organic compounds when analysed via laboratory GC-MS.	• vinyl-chloride results may vary depending on sample gas volume injected into tube.	**Methodology** Autothermal desorption tubes are steel cylinders typically 10 cm long containing a solid polymer adsorbant which has the ability to absorb gases from the surrounding atmosphere over a known time period or from a known volume of gas sample injected into the tube. The gas absorbed is fixed within the polymer and after the sampling period the tube can be sealed with end-caps and sent to a laboratory for analysis of the desorbed gas mixture (achieved by heating the tube prior to GC-MS analysis). **Application** The Environment Agency's LFTGN 04 method for analysing landfill gas trace compounds, and sampling ambient or workplace air quality. Different types of adsorbent media are available for different chemical groups, detailed in EA Technical Report P1-438 *Investigation of the composition and emissions of trace components in landfill gas* (EA, 2002). Such tubes may be used when indications of soil volatile organic contamination has been found, for example by semi-quantitative PID headspace/borehole sampling: ATD tubes provide full speciation of the VOC mixture.
Portable laser methane detection	• specific to methane.	• can sample up to 30 m away from methane source and can read through window glass • portable, light, very rapid response time • external battery packs allow eight hour operation • no interferences. Reads on parts per million range (ppm).	• care needed in data interpretation • readings are in ppm per metre of pathlength ie a 1 m pathlength of 100 ppm gives the same reading as a 0.5 m pathlength of 200 ppm • reading is the average over that distance • internal battery life limited (three hours) • not designed to be intrinsically safe.	**Methodology** The laser radiation is absorbed by methane molecules at specific frequencies and the degree of absorption in backscattered radiation is measured. **Application** Remote gas leak detection; internal monitoring of buildings after safe atmosphere has been proven using appropriate EU ATEX Directive certified intrinsically safe equipment (warning: UK Government Health and Safety Executive confined space entry requirements must be met. See <www.hse.gov.uk/pubns/indg258.pdf> and related documents). Post construction monitoring of subfloors voids.

Note: All pictures are for illustrative purposes only

5.2 SELECTION PROCESS

Before the monitoring instrumentation is selected for the site, a number of factors need to be considered. These factors taken from CIRIA publication R131 (Crowhurst *et al*, 1993) are listed below. Consideration of these questions will enable the most suitable soil gas monitoring instrumentation to be selected:

1. *Is the instrument intrinsically safe?*
2. *What do I want to measure?*
3. *Can I take the measurements myself?*
4. *Can all measurements required be made with a single instrument?*
5. *Is the instrument robust?*
6. *Does the instrument have an independent power supply?*
7. *How heavy and what size is the instrument?*
8. *Does it have its own sampling pump?*
9. *What type of display system is used?*
10. *Are the operating instructions clear and precise?*
11. *Are the controls clear and easy to use?*
12. *Does the instrument have an internal memory and computer logging facility?*
13. *With what mixture was the instrument calibrated?*
14. *How frequently should the instrument be serviced and/or calibrated and what is the cost?*
15. *Does the instrument's sensor have a limited normal life?*
16. *Could the life of the sensor be reduced by components of the sampled environment?*
17. *What customer support does the manufacturer offer?*

Once the instrument has been selected, it is vital that all calibration processes for that particular piece of equipment are undertaken. Typically, the instrument needs to be calibrated before leaving the office/working area, and again on return from the monitoring round. In some circumstances (for example extended periods on site) it will be necessary for this pattern of calibration to be carried out on site/out of the office/working area.

5.3 MONITORING METHODOLOGIES

A monitoring methodology/protocol should be followed consistently during each round of soil gas monitoring. The consistent application of methodology is essential to ensure comparable results are obtained from different monitoring rounds. On many occasions, soil gas monitoring at a site will be carried out by different operators. The adoption of a well-defined monitoring methodology/protocol will minimise the potential for error/anomaly arising from different operator practice.

5.3.1 Soil gas monitoring

Previous guidance (Card, 1996) states that as a minimum, the methodology for soil gas monitoring should include the recording of the sampling point, date, time, weather conditions and the depth of the water table within the well. Ideally the monitoring methodology should incorporate checklists, typically of required equipment and data, and instructions to follow. This general methodology may then be adapted for site-specific circumstances, and then described and incorporated as part of the final report.

All monitoring results/information should be recorded on a separate proforma for each monitoring round/visit. This will aid quality assurance and data interpretation. An example of a proforma is included as Appendix A3. The proforma is set out to allow the operator to collect the ambient and site-specific data, such as the site name and weather conditions, before the soil gas monitoring. A table is also provided to ensure that all the required data is collected in the correct order. For example, the flow rate and pressure are required before the soil gas concentrations. It is good practice to enclose each completed proforma in date order into the report.

Examples of sampling methodologies/protocols are provided in CIRIA publication R131 *The measurement of methane and other gases from the ground* (Crowhurst *et al*, 1993) and the Environment Agency's guidance *The measurement of trace components of landfill gas* (Environment Agency, 2004b). A third example has been included in Example 5.1 which provides good practice guidance for monitoring of boreholes using a commonly used flow meter, an instrument containing an infra-red sensor and dip meter. The example consists of a series of steps (Crowhurst *et al*, 1993).

Historically, gas composition was the only measurement routinely made during soil gas monitoring. While this measurement is one of the most important due to its governing influence on the risk of asphyxiation, explosion and toxicity, there are other properties that must be measured to properly determine the soil gas regime on a site (see Table 5.2).

Table 5.2 *Recommended data collection in monitoring programmes (Boyle and Witherington, 2007)*

Data	Reason for collection
Concentrations of components within soil gas (methane, carbon dioxide, oxygen etc)	The concentration of soil gas within the installation is a significant determinant of risk.
Pressure of gas	Increased soil gas pressure within the borehole can cause horizontal or vertical movement of gases. As a result, it can greatly influence the soil gas migration pathways.
Gas flow	The gas flow measurements (volume of gas being emitted from a monitoring well per unit time) enable the surface emissions (volume of gas escaping from a unit area of ground in a unit of time) to be estimated which in turn can be used to calculate the potential for gas ingress into a proposed building.
Volume of gas contained in the monitoring structure	It is common to get an initial high reading (peak) followed by a lower reading. This suggests that a small volume of higher concentration gas is present. Both the peak and steady reading should be recorded.
Potential depletion of gas during sampling	Monitoring will deplete the volume of gas in the borehole and if low volumes of gas are present the monitor reading will reduce. To estimate the reservoir of gas in the soil a pumping trial can be carried out. For this, gas will be pumped out of a borehole for a measured period of time and the return of the gas can be monitored.
Water level within monitoring well	The groundwater level is typically routinely recorded during the monitoring. However, the significance of the groundwater is not always fully appreciated. If the water table is particularly high or varies greatly, there is the potential for the water to rise above the response zone. This can prevent the ingress of soil gas into the well. The lack of concentrations in the well can subsequently give false readings. A rising groundwater level can cause an increase in gas concentration. Rising groundwater will reduce the pore space in which the soil gas exists. The soil gas will be displaced upwards causing the soil gas to migrate upwards and be released from the ground into the atmosphere, or into buildings or boreholes (the piston effect).
Tidal variations	Changes in tide levels in coastal areas result in rises and falls in the water table. As discussed above, these can affect soil gas ingress and increase the potential for migration or ingress into buildings (the piston effect).
Depth of response zone	Recording the response zone depth confirms the strata and area of contamination soil gases and/or vapours are being monitored from.
Total depth of monitoring well	To ensure no blockages are present in the well and to confirm response zone.
Atmospheric pressure and rate of change	Atmospheric pressure can significantly affect the rate of gas release. Pressure changes can affect the gas expansion or reduction rate and hence the concentration measurement. A monitoring programme should always endeavour to include one or more sets of data obtained during falling/low pressure atmospheric conditions.
Ambient temperature	Varying temperatures can have a small impact on the production of soil gas, from biological activity.
Date, time, site details, sample location	It is standard good practice to record the date, time and site details for each monitoring visit. This will ensure that the data can be interpreted by another operator and at a later date.
Weather conditions and recent weather	Rainfall can affect the groundwater which in turn affects soil gas ingress. A high surface wind speed can influence the surface emission rates, which can lead to lower borehole concentrations. It is recommended that the weather conditions are recorded approximately 24 hours before completing the monitoring. This information can easily be found on any weather or news dedicated website <www.news.bbc.co.uk>. This information will be aid in interpreting the results at a later date.
Condition of monitoring point	Comments on the monitoring point can relate to either the gas tap or the immediate surrounding surface area or both. Observations such as 'an open gas tap' or 'blocked gas tap' can be used to later interpret the data.
Ground conditions	The ground conditions on a site can affect the gas regime. For example, ground frost and flooded ground can trap the soil gas beneath the surface by reducing the soil pore space available for the soil gas to release to atmosphere. This can increase the potential for lateral migration through soils.
Vegetation stress	Vegetation stress can be an indication of soil gas beneath the surface. Recording any signs of vegetation stress, not only adds to the evidence but also can be used to identify migration pathways.
Visible, audible or olfactory signs of gas migration	As above, any recorded observations of strong odours, visible oil staining and discolouration can be used to later interpret the data. If odours are noted/recorded during routine monitoring it is recommended that monitoring/sampling/chemical analysis for trace gases should be undertaken.

Example 5.1 Typical gas monitoring round

1. Record daily (and if appropriate hourly) atmospheric pressure readings during period before the monitoring visit.

2. Calibrate the instruments before the monitoring visit.

3. Before starting the monitoring, turn on the monitoring equipment, attach tubing and run through clean air and zero the methane. This needs to be done well away from any sources of soil gases and/or vapours such as vehicles and monitoring locations.

4. Keep the monitoring equipment switched on between boreholes to prevent having to zero the methane each time it is switched on. However, ensure methane is zeroed before beginning to monitor at subsequent wells. Keeping the monitoring equipment on also purges any residual gas.

5. Record the atmospheric pressure reading from the monitoring equipment. Also record weather, record air temperature and ground condition at the site. This information is important as it may influence the interpretation of the gas results.

6. Switch on the flow meter, attach the inlet tube to the gas tap and open. Record the range of pressures and flow readings on the gas monitoring proforma, making sure "positive" or "negative" is recorded.

7. Close the gas tap and remove the gas flow meter.

8. Attach the monitoring equipment tubing to the gas tap and open. Switch on the pump and record the **peak and steady** reading for methane (% v/v), methane (% LEL), carbon dioxide (% v/v) and oxygen (% v/v). It is also good practice to record the time taken to reach the steady reading.

9. If the gas readings have not reached a steady value after three minutes, record the concentrations and the direction and rate of change in concentration (that is steadily increasing/rapid decrease). Where the concentrations are decreasing always record the peak concentration.

If very high readings are recorded on the monitoring equipment it is worth monitoring the well for a longer period (up to 10 minutes) to determine if the concentrations are related to build up of gas in the well (for example from a pocket of methane within a layer of alluvium) or are being constantly replenished by methane or carbon dioxide from the soil. The readings over time should be recorded on the gas monitoring sheets.

Note: The Monitoring equipment is liable to suck up water from the boreholes. Watch the clear plastic tubing (attached to the tap) carefully and if this should happen, quickly detach the tubing from the inlet and switch off the pump. Record the gas concentrations and make a note that water was sucked up. Check the filter and if wet, replace with a dry filter. Remove water from the tubing.

10. Once data is recorded remove the tubing from the gas tap and close the tap. Purge the monitoring equipment in clean air (away from the borehole/and other sources of gas) until the methane and carbon dioxide concentrations return to zero and the oxygen is reading atmospheric concentrations.

11. Record the water level using a dip meter, usually obtained by removing the gas tap or cover from the borehole. Water level readings are usually recorded from the top of the borehole or from ground level or both (be consistent and note to where depth relates), the top of the water. After obtaining a reading, record on the proforma and replace the gas tap or cover ensuring that the tap is closed and cover locked.

12. Make a note of any defects to the boreholes and perform maintenance if appropriate.

13. Repeat for all boreholes and record an atmospheric pressure reading once all monitoring has ceased and record on proforma. Note any trend in atmospheric pressure in the lead up to and during the monitoring visit.

Note: The steps listed in Example 5.1 is an example and does not represent generic advice

5.3.2 Specialist monitoring

Once the initial data has been gathered and recorded it may be necessary to use different monitoring techniques to further understand the nature of the gas source, particularly if high or variable gas concentrations are detected (for example peat layer with high methane concentrations). There are three different approaches that can be considered:

1   Recirculation.
2   Venting.
3   Pumping to sample container.

These monitoring techniques can provide additional information about the gas regime which can be used to support the risk assessment. However, these techniques have not been used widely and further research is necessary to fully understand the results that they produce (see Table 5.2).

Recirculation

Recirculation is undertaken to measure the rate of gas recovery in a borehole. This consists of flushing a monitoring well with nitrogen or other inert gas, such as helium from a gas cylinder. The gas stream disperses into the soil gas in the well. A gas analyser is then used to monitor the rate at which the soil gas returns to the standpipe. The rate of change in soil gas concentration with time represents the rate of equilibrium of soil gas in the monitoring well with the surroundings due to diffusion and convection (Raybould *et al*, 1995).

Flushing is not commonly undertaken because of concerns over the accuracy of the results, the potential for error and concerns over the potential for spontaneous combustion. For example, to undertake such monitoring the volume of the standpipe is required. However, there is a potential for open voids to be present surrounding the standpipe (due to either natural strata characteristics or disturbance during drilling). The presence of open voids within the vicinity of the standpipe, which are generally impossible to determine, will increase the volume of soil gas collected. Assumptions can be made that the volume calculated relates only to the monitoring well and excludes any unknown open voids. However, this assumption can result in significant errors in the calculation of soil gas equilibrium (Raybould *et al*, 1995).

Earlier guidance, CIRIA publication R131 (Crowhurst *et al*, 1993) commented that recirculation as monitoring technique should be used only where the emission rate is very low or where there is no flow or positive pressure within the monitoring well. Concerns still remain with respect to several areas of complexity including the potential for spontaneous combustion, the need for intrinsically safe instrumentation and equipment, together with the accuracy/meaning of results obtained (see Example 5.1). Further research is required on the technique of recirculation.

> **QUICK REFERENCE Recirculation considerations**
>
> When undertaking recirculation methods further considerations should be given to:
>
> - the use of air as the purging gas, which can potentially result in spontaneous combustion within the body of fill, is not recommended without first establishing a sound understanding of the site/waste conditions, the amount of air introduced etc
> - the use of gas monitoring instruments can create a negative pressure in the borehole, affecting the rate of equilibrium. This effect can be minimised by recirculating the gas back into the monitoring well (provided two gas taps are present) (Godson et al, 1996)
> - the use of inappropriate gas monitoring instruments which can alter or destroy the gas composition before recirculation back into the monitoring well (see Table 5.1).

Venting

It is common practice to close the gas taps in between monitoring rounds. However, by leaving the gas tap open it allows the soil gas in the borehole to vent to atmosphere which is thought to represent a "realistic" situation. Venting can also convert anaerobic conditions to aerobic conditions, hence reducing the soil gas production artificially. There is very limited experience and data gathered for venting, and further research is required into this technique (see also recommendation 1 of Chapter 11).

Pumping to sample containers

This technique includes venting the purged soil gas direct into sample containers. This monitoring/sampling technique may be used to confirm the soil gas concentrations recorded by the monitoring instrument with the use of laboratory analysis. In addition, the technique may be beneficial where limited soil gas is present in the standpipe available for sampling. The use of soil gas monitoring instruments for sampling is restricted to those such as the infra-red gas analyser that do not destroy the sample.

Table 5.3   *Summary of specialist monitoring techniques*

Monitoring technique	Advantages	Disadvantages	When to use
Recirculation	- Identification layering of gas within the bore hole - allows recirculation where limited gas sample present.	- requires 50 mm diameter well minimum - internal tube can flood and draw water up into gas monitoring more easily than normal gas taps.	
Venting to atmosphere	- may be more representative of actual soil pore gas concentration.	- the open gas tap may provide a preferential path flow - potential to provide a soil gas concentration that is not representative of gas ingress to buildings - convert anaerobic conditions to aerobic.	It is not recommended to use this monitoring technique until further research into the advantages, interpretation of the results and any health and safety issues are undertaken.
Pumping to sample containers	- allows collection of sample at same time as monitoring to allow confirmation of the measurement - indicates gas concentration of sampled gas during monitoring - requires less gas sample from the standpipe as separate monitoring and sampling techniques.	- instruments that alter or destroy the measured gases cannot be used with pumping.	This technique should be considered when limited gas sample is present in the standpipe however the use of gas monitoring instruments needs careful consideration.

5.3.3 Headspace analysis

Not all soil gas monitoring has to be carried out with the use of installation wells. For example, the "headspace procedure" is a quick and simple field screening procedure used to determine the presence of volatile organic compounds in soil or water, before a full site assessment is conducted or during a Phase II site investigation.

> **QUICK REFERENCE Field headspace test**
>
> The "headspace procedure" involves collecting a soil or water sample, placing it in an air-tight container and then analysing the headspace vapour using an appropriate portable analytical instrument (eg a PID or FID depending on the gas of interest). The "headspace" is the area between the sample and the top of the container. This method is based on the US and Canadian standard methodology, but as a modification, zip lock plastic bags can be used instead of jars, provided they are sealed tightly.

As with visual observations, headspace concentrations can aid in identifying appropriate locations for soil gas installations as well as appropriate response zones during the ground investigation. Typically, cost restrictions mean that not all exploratory locations are installed with soil gas monitoring wells. However, consideration of headspace monitoring can aid in locating the wells across the site in the areas where soil gas is most likely to occur. The monitoring can also assist the design of the vertical response zone by determining the stratum from which the higher concentrations were detected.

The following factors need to be considered when determining headspace:

- the headspace needs a minimum of 20 minutes to develop
- ambient temperatures and weather conditions during headspace analysis should be recorded
- samples should not be left in direct sunlight
- headspace analysis should be completed on the same working day that the sample is collected
- correlation with visual/olfactory evidence (that is the soil description) is essential
- headspace readings and depth samples taken should be recorded on the exploratory hole logs.

5.3.4 Internal gas survey

Internal gas surveys are generally undertaken within existing buildings, structures and service ducts where there is a potential for soil gas ingress. An example of an internal gas survey methodology for detecting the ingress of methane using an FID is described in Example 5.2.

> **Example 5.2 Internal gas survey with FID**
>
> 1. Before starting the monitoring, record details of the property and note any previously unrecorded details such as extensions/conservatories.
> 2. Turn on the monitoring equipment, attach probe and run through clean air to zero. This needs to be done well away from any sources of soil gases and/or vapours.
> 3. Record weather, air temperature and ground condition at the site.
> 4. Sketch a ground floor layout plan or individual room layout plans with dimensions.
> 5. Survey one room at a time. First stand in the centre of the room with the probe at head height, measure the ambient concentration and record on the room plan.
> 6. Next measure any soil gases and/or vapours ingress that could enter the room from beneath the floor. Place the probe at floor level where the floor meets the wall (where skirting boards are present monitor above and below the skirting board) and very slowly walk around the room watching the meter. Where readings are above ambient, record the concentration and location on the layout plan.
> 7. For floors that are bare or boarded, walk across the floor and measure along cracks. Other sources such as carpets and house cleaning materials can also give off VOCs.
> 8. Soil gases and/or vapours can accumulate in confined spaces such as cupboards, so note locations and measure inside all cupboards. Do not enter confined spaces if the ambient concentrations exceed 1000 ppm.
> 9. The main points of entry for gas to enter buildings are through service entries and around and beneath the skirting board (see Figure 2.1). Therefore, locate, note and measure around all service entries and skirting boards, commenting on whether they enter through the floors or above ground.
> 10. Elevated readings can occur when people are smoking, cooking, cleaning etc. Therefore note any such activities or odours on each room plan.
> 11. Repeat for all rooms.
> 12. Once all the ground floor and basement areas have been surveyed. Walk around the external perimeter of the building and note and measure around any airbricks, manhole covers, service entries, shallow excavations etc.
>
> **Notes:**
>
> *As with measurements of gas in the ground, measurements of gas in buildings is also subject to variability (eg temporal and seasonal variations and activities of occupants etc).*
>
> *As stated in Table 5.1, some models of FID are not intrinsically safe in confined spaces. It is recommended that an intrinsically safe instrument is used to undertake the survey.*
>
> *The accumulation of dust in the tubing/filter can cause the FID meter to record false readings. If elevated readings above zero are recorded in clean external air, filters and tubing should be checked and replaced as necessary.*

Note: The steps listed in Example 5.2 are an example only.

5.3.5 Quality assurance

Completion of a proforma for each monitoring visit for all types of monitoring is part of the recommended quality assurance procedures.

5.4 FLUX BOX MEASUREMENTS

Flux chambers (also known as flux boxes) comprise a box that is open in the base that is placed over a gassing site, allowing gas that is being emitted from the surface to be collected (Figure 5.1). The rate of gas accumulation is measured at timed intervals with a watch, using an FID connected to a sampling T-bar within the chamber. The surface emission rate is then calculated. Flux boxes are considered by some practitioners to be of most value in circumstances of low gas emission rates, typically caused mainly (but not exclusively) by diffusion processes.

Note: Emission is a combination of pressure-driven flow (eg brought about by changes in atmospheric pressure) and diffusion (differences in concentration).

Flux chambers measure surface emission rates only and can be used to give more confidence in emission rates estimated from borehole data. However surface conditions might prevent any emission even in "active sites". They do not give any indication of the levels of gas in the ground or generation rates. Data from flux chambers should not be used in isolation for gas risk assessments/gas protection design for proposed existing development sites.

Comprehensive guidance is provided by the Environment Agency on measuring surface emissions using flux chambers (Environment Agency, 2004a). However this guidance was specifically developed for compliance monitoring of landfill caps and the analysis method reflects this. The methodology is appropriate for development sites but other methods of analysis may be required. Practical limits of detection of methane using flux chambers are from 5×10^{-5} mg/m^2/s to 5 mg/m^2/s. The Environment Agency advise that flux chamber readings should not be undertaken when the ground is saturated or covered by frost and, ideally, such tests should be undertaken when barometric pressure is not at high or low extremes (that is when rate of change may be very small) or when wind speeds are greater than 3 m/s (Beaufort Force 2) as the readings may be adversely affected.

There are two types of flux chambers:

- static (or closed system) – where the gas is simply allowed to accumulate in a closed chamber
- dynamic (or open system) – where controlled positive flow of fresh air is passed through the chamber (for example via a pump or a pressure relief valve).

Many boxes that are perceived to be static are actually dynamic, because some instruments used to monitor for gas remove air from the chamber. This can lead to significant errors in assessing the results (see Table 5.4).

Table 5.4 *Summary of advantages and disadvantages of static/dynamic chambers*

	Static chamber	Dynamic chamber
Advantages	• can be left in place to monitor over a long period.	• dynamic chambers have less effect on the soil gas regime because there is no induced draw or restricting effect on the soil gas emissions • can obtain several results in one day • data obtained on site.
Disadvantages	• off-site analysis of collected samples required • it has been recorded that static chambers do not give a reliable indication of surface emissions because they either restrict movement of gas into the chamber as gas pressure builds up or they enhance gas flows as suctions develop due to removal of air during sampling – this should be referenced inline with other entries.	• impractical to leave on most sites for a long period.

Figure 5.1 *Flux box*

5.5 NUMBER, FREQUENCY AND DURATION OF MONITORING

5.5.1 Number of monitoring rounds

Previous guidance suggests that a minimum of six to 10 readings are required before any confident decisions can be made about a gas regime (Harries *et al*, 1995). This advice has been reiterated in other more recent guidance (Environment Agency *et al*, 2000) and also appears to be the minimum that is acceptable to the majority of regulators assessing development sites. **However, focusing on a minimum number and period of readings can be misleading and the key should be that the monitoring period for a specific site covers the "worst case" scenario.** Such a "worst case" scenario will occur during falling atmospheric pressure and, in particular, weather conditions such as rainfall, frost and dry weather. Other site-specific conditions (for example tidal) can also affect the gas regime in the ground (Raybould *et al*, 1995). It should be noted that most rapid falls in atmospheric pressure occur when the pressure was initially high, for example 1010 or 1020 mb.

There is a balance to be considered between the cost of additional monitoring and the improvement in technical confidence which may result. Again, phasing or staging of a monitoring programme for a particular site can play a crucial role. Consideration should also be given to the overall objectives of, and context for the assessment (Environment Agency *et al*, 2000). The benefits of the additional information and whether it is likely to change the scope of gas protection is also important, as are the consequences of failing to characterise adequately pollutant linkages. Investigations concerned with soil gas are required to provide monitoring data sufficient to allow prediction of worst case conditions enabling the confident assessment of risk and subsequent design of appropriate gas protection schemes. Monitoring programmes should not be an academic exercise in data collection. However, the corollary to this is that a small data set will result in reduced confidence of the risk assessment and an increased scope of gas protection to provide additional factors of safety/levels of confidence. This is unlikely to be the case on sites with highly sensitive proposed end uses, or high gas generation potential.

For sites of low generation potential, which consistently record low concentrations of soil gas under worst case conditions, a limited programme of monitoring (in terms of both frequency and duration) is likely to be appropriate. However, for those sites where high or variable concentrations are recorded, extended periods of monitoring and possibly the installation of additional gas monitoring wells may be necessary. Using this philosophy a matrix has been developed that will aid in determining an appropriate number of gas monitoring visits (see Tables 5.5a and 5.5b).

Table 5.5a   *Typical/idealised periods of monitoring (after Wilson et al, 2005)*

		Generation potential of source [2]				
		Very low	Low	Moderate	High	Very high
Sensitivity of development	Low (commercial)	1 month	2 months	3 months	6 months	12 months
	Moderate (flats)	2 months	3 months	6 months	12 months	24 months
	High (residential with gardens)	3 months[1]	6 months	6 months	12[1] months	24 months

Table 5.5b   *Typical/idealised frequency of monitoring (after Wilson et al, 2005)[3]*

		Generation potential of source [2]				
		Very low	Low	Moderate	High	Very high
Sensitivity of development	Low (commercial)	4	6	6	12	12
	Moderate (flats)	6	6	9	12	24
	High (residential with gardens)	6[1]	9	12	24[1]	24

Note

1   NHBC guidance also recommends this period of monitoring (Boyle and Witherington, 2007).

2   There is no industry consent over "high", "medium" or "low" generation potential of source.

3   At least two sets of readings should be at low and falling atmospheric pressure (but not restricted to periods below <1000 mb) known as worst case conditions. Historical data can be used as part of the data set (Table 5.5b).

Not all sites will require gas monitoring for the period frequency indicated in Tables 5.5a and 5.5b. Depending on specific circumstances, additional readings may be required However, this would need to be confirmed with demonstrable evidence. NHBC's publication, *Guidance on evaluation of development proposals on sites where methane and carbon dioxide are present* recommends that extended period of monitoring and possibly the installation of additional gas monitoring wells may be needed on sites where consistently record high or variable concentration (Boyle and Witherington 2007).

It should be noted that during longer periods of monitoring, once the temporal pattern of gas emission is established, monitoring frequency can be decreased unless specified threshold concentrations are detected (see Chapter 7).

Consideration should be given to the equilibrium of the soil gas regime while assessing monitoring data. It is recommended that newly-installed monitoring wells are left for 24 hours to allow the soil gas to reach equilibrium. It should be recognised, however, that some soil gas regimes could take considerably longer (up to seven days) to reach equilibrium (eg low generation, low permeability situations). In such circumstances, interpretation of any initial readings should take this equilibrium process into account.

> **Case study 5.1 Example of typical gas monitoring round**
>
> **The need for a flexible monitoring programme and long-term monitoring where there is a lack of Phase I desk study information**
>
> At one particular site there was no obvious source of gas (the ground conditions were natural London Clay formation) but monitoring in one well shortly after installation revealed up to 10 % v/v methane. The monitoring regime was extended and the frequency of visits increased. The results showed that early results were not representative of the site and were caused by disturbing activities when the well was installed. The increased frequency and extended period of monitoring demonstrated a gradual but consistent decrease in concentration to an equilibrium value that was less than 1 % v/v.
>
> **Example of a poorly-designed soil gas monitoring programme and the possible implications of such a programme**
>
> A ground investigation was submitted containing gas-monitoring information. Three monitoring events had been undertaken, all during the week immediately following installation of the wells, and during a period of high atmospheric pressure. The local authority determined that this was inadequate, and the monitoring immediately following well installation was inappropriate, as the soil gas regime had not had sufficient time to reach equilibrium. Additional monitoring was undertaken which eventually demonstrated that the soil gas regime on the site did not present a potential risk to the development requiring mitigation. However, the development had been delayed for the consultation period with the local authority, while the additional monitoring occurred.

5.6 PRESENTATION OF DATA

As a minimum, presentation of gas data in reports should include the following:

- site plan (showing monitoring locations, identifying site zones, source areas etc)
- raw data (usually set out in proformas) see Appendix A3
- summarised data (set out in tables/charts) see Table 5.6.

It is also essential that all of this data is set in the context of the site itself. The conceptual site model should be adequately described in the text and illustrated by appropriate plans, drawings, cross-sections etc.

Presentation of the factual gas monitoring data should demonstrate that the monitoring/sampling was carried out consistently following appropriate quality assurance procedures. The summary of the raw data provides a clear overview of the results, showing trends (for example often aided by graphical presentation or charts). The text interpretation of the results then puts the values into context.

If a factual report is required only the raw data should be presented including plans and cross-sections together with proformas or summary tables/charts. A summary table alone (that is without the supporting data) is not sufficient in any type of report.

Table 5.6   *Idealised example: Summary of monitoring data*

Borehole	Response zone/strata	Evidence of contamination	No. of monitoring occasions	Methane (%)	Carbon dioxide (%)	Oxygen (%)	Flow (l/hr)	Water levels (meter below ground level)	Range of atmospheric pressures during monitoring round (Pa)
BH1	1–6m/W	No	6	16.2–22.3	8.4–12.1	11.7–15.6	0.5–1.8	4.81–5.20	996–1020
BH2	1–3.5m/MG	Hydrocarbon odour 2.0–3.0 m	6	80.4–100	0.5–1.2	5.3–8.9	<0.01	Dry	996–1020
BH3	1–4.0m/MG	No	6	0–0.9	0.5–2.2	17.3–21.0	<0.01	Dry	996–1020
BH4	5.6–10.0/C	No	6	0	2.0–2.5	19.8–20.5	<0.01	6.69–6.72	996–1020
BH5	3.5–4.2m/P	Organic odour	6	42.3–55.8	2.3–4.4	3.2–4.8	0.05–0.08	Dry	996–1020

W = Waste, MG = Made Ground, P = Peat, C = Chalk

Table 5.6 illustrates that by installing response zones within specific stratum and by recording all data and information available, the first stage of interpretation may be undertaken with appropriate cross-reference to the locations of the boreholes on site, and any significant features deriving from the natural geological strata and the nature of the infilled ground.

The following patterns/trends are illustrated from the data above:

- high methane and carbon dioxide concentrations have been detected in the waste on all six occasions and the oxygen concentrations are also depleted
- high methane concentrations, slightly elevated carbon dioxide and depleted oxygen has been recorded within the made ground at one location. However, it has been noted that a hydrocarbon odour was detected at this same location. Due to absence of methane concentrations at the second made ground monitoring well, it is shown that the high concentrations are not widespread within the made ground, but may be associated with a hydrocarbon spill
- within the natural ground, only slightly elevated concentrations were recorded in the chalk. However, within the peat, an organic odour was noted and associated methane and carbon dioxide concentrations detected. These concentrations appear to be localised due to the low concentrations in the made ground overlying the natural stratum
- the highest flow readings were recorded in the waste and the peat.

It should be noted that the recorded groundwater levels are also relevant. The majority of the wells are dry or the groundwater levels are low and below the response zones, but at BH4, the groundwater had entered the response zone. There is still an unsaturated zone above the groundwater. However, if the groundwater rises there is a potential for the response zone to be saturated (flooded) and inhibit gas flow into the monitoring well.

5.7 SUMMARY

1. A good understanding of equipment operation will ensure the correct instrument is utilised and all limitations and potential interferences are understood.

2. Monitoring instruments must be calibrated before and after monitoring.

3. The adoption of a well-defined monitoring protocol will ensure consistency.

4. As a minimum, the following measurements should always be taken during each monitoring round:
    i. Gas composition.
    ii. Gas flow rate.
    iii. Water level.
    iv. Atmospheric pressure.

5. Monitoring should, as far as practical, be undertaken during times of falling atmospheric pressure and various weather/site-specific conditions (rainfall, frost and tidal influences) to ensure data is acquired during "worst case" conditions.

6 Sampling methodologies

6.1 SAMPLING FOR LABORATORY ANALYSIS

6.1.1 Introduction

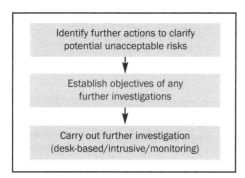

Note: Extracted from Figure 1.1

Gas sampling and analysis is relatively inexpensive, rapid and will add value to any soil gas investigation. The analysis can confirm the nature and source of the ground gases confirm field measurements and quantify hazardous/odorous trace constituents. The adoption of a well-defined monitoring protocol including laboratory analysis will ensure consistency.

This chapter describes the available gas sampling techniques and the subsequent laboratory analysis, with additional detail in Appendix A4.

6.1.2 Bulk gas sampling

There are two common sampling techniques for bulk gases:

1. Pressurised sampling cylinders
2. Non-pressurised sampling vessels.

CIRIA publication R131 (Crowhurst *et al*, 1993) and the more recent EA guidance (Environment Agency, 2004b) provide detailed descriptions of the techniques listed above and the factors which can influence the sample. In addition, the specialised laboratories are also a source of information and advice.

6.1.3 Pressurised sampling cylinders

A pressurised sampling cylinder typically comprises a hand pump, (an example of which is a Gresham Pump) used to compress the sample of gas into a small aluminium or stainless steel cylinder. The cylinders which are available in various sizes from 15 ml to 110 ml include a valve at each end (Crowhurst *et al*, 1993). Once the sample has been collected in the cylinder, it can be sent for further analysis.

There are a number of pressurised sampling cylinders available, which have been listed in Table 6.1. The advantages and disadvantages of using such techniques are also discussed.

Table 6.1   *Summary of pressurised sampling vessels (based on Institutes of Waste Management, 1998)*

Container	Advantages	Disadvantages	When to use
Stainless steel bombs (typically used with Gresham Pump)	• robust • relatively small size • pressurised contents prevent condensation in vessel • pressure prevents ingress of ambient air • carbon steel bombs can be used to collect gas samples for bulk constituents analysis.	• relatively heavy • body may be reactive to some gases such as hydrogen sulphide/mercaptans and sulphur compounds • previous samples may leave traces on valves and body • if pressure lost, sample is lost • more expensive than bags.	Suitable for bulk gases and trace components especially odorous/nuisance compounds.
Copper and aluminium	• robust • pressurised versions available.	• more reactive than stainless steel • relatively heavy • permeable to hydrogen.	Suitable for bulk gases and trace components.
Glass	• relatively inert • easy visual inspection • easily purged.	• fragile • problems with seals.	Glass sampling bulbs are not recommended simply due to the fragile condition of the containers. Damage is possible both in the field and during transportation.

6.1.4 Non-pressurised gas sampling vessel

A non-pressurised gas sampling vessel consists of a vessel sealed at both ends by taps or valves. The vessel, which is connected to the sampling point, is also attached to a vacuum or hand pump. A common vessel is the tedlar bag. Other common vessels, along with their advantages and limitations, are summarised in Table 6.2 below.

Table 6.2   *Summary of non-pressurised sampling vessels (based on Institute of Wastes Management, 1998)*

Container	Advantages	Disadvantages	When to use
Tedlar bags	• cross contamination avoided • lightweight • less likely to suffer pressure loss • non contaminating material used.	• fragile, easily punctured, hard to store safely when full.	Suitable for bulk gases and vapours. Layered or black tedlar bags will be required for the collection of VOCs to prevent light degradation. Recommended holding time for non reactive gases is 72–120 hours.
Teflon bags	• large capacity available – up to 200 litres • lightweight • chemically inert • non-porous to hydrogen.	• fragile • vacuum cleaning required • water vapour condenses inside bag.	Suitable for bulk gases and vapours. However, not recommended for reactive compounds, such as hydrogen sulphide. Recommended holding time for non reactive gases is 72–120 hours.
Polyester/vinyl bags and aluminium bags	• large capacity available – up to 44 litres • lightweight • inexpensive.	• fragile, easily punctured, hard to store safely when full • vacuum cleaning required • water vapour condenses on inside and hard to clean out • body material may react to some minor components • permeable to hydrogen.	Suitable for bulk gases and vapours. However, not recommended for reactive compounds, such as hydrogen sulphide. Recommended holding time for non reactive gases is 72–120 hours.
Rubber bags	• large capacity available • lightweight.	• fragile and easily punctured • may react with sample • water vapour condenses in bag.	Suitable for bulk gases and vapours. However, not recommended for reactive compounds, such as hydrogen sulphide. Recommended holding time for non reactive gases is 72–120 hours.
Kevlar bags	• large size available • lightweight • relatively inert.	• fragile and easily punctured • may react with sample • water vapour condenses in bag.	Suitable for bulk gases and vapours. However, not recommended for reactive compounds, such as hydrogen sulphide. Recommended holding time for non reactive gases is 72–120 hours.

6.1.5 Trace gas sampling

The sampling techniques described above are typically used for the collection of bulk gases. However, such instruments are not normally suitable for the collection of trace gases and absorbent columns/cold traps are more suitable for collecting such trace components.

Absorbent columns/cold traps are based on a high-activity sorbent used to collect the gas. The sample is then transported to the laboratory where the trapped chemicals are solvent – extracted or thermally desorbed. This technique can concentrate particular groups of a compound and is also suitable for detailed analysis of trace components. However it has not been widely used in the industry and requires a high level of expertise. The use of such a technique is recommended for the collection of trace gases in order to minimise cross contamination.

The recent EA publication *Guidance for monitoring trace components in landfill gas* (Environment Agency, 2004b), contains further detail of specific absorbent materials for the priority trace components.

6.2 METHODOLOGY

As with soil gas monitoring, sampling methodologies need to ensure representative samples are collected. Analysis of the sampled soil gas and vapour will:

- determine its chemical nature
- help identify the source
- confirm the identity of any odorous and toxic constituents
- confirm field measurements.

Measuring a representative soil gas concentration immediately following installation of gas wells can be difficult. This is because the disturbance to the ground during the excavation of the borehole and installation of the monitoring well can affect the gas regime. Sampling of soil gas should not take place immediately following borehole installation.

Techniques available for collecting a gas sample are described earlier. Regardless of the technique used, it is recommended that a methodology/protocol is followed consistently during all gas sampling. CIRIA publication R131 (Crowhurst *et al*, 1993) provides an example of a methodology for soil gas sampling. The more recent EA guidance (Environment Agency, 2004a) also provides a general monitoring and sampling methodology. By following such a methodology the operator(s) will collect samples in a consistent and efficient manner. Such consistency is essential to ensure appropriate interpretation. An example methodology for the sampling procedure using a Gresham Pump is presented in Example 6.1.

> **Example 6.1 Use of a Gresham pump in gas sampling**
>
> 1. The bulk gas sample is taken using a Gresham Pump and a clean cylinder. Sampling should be carried out immediately following the monitoring of a borehole. First check the cylinder is pressurised by attaching it to the pump. If the needle is half way in the green it is OK. If not, pressure has been lost since the last time it was pressurised and the tube should not be used.
> 2. If sufficiently pressurised, release the pressure by depressing the pin at one end.
> 3. Place cylinder on the pump and check that pressure has been fully released.
> 4. Attach tube to gas tap and open. Begin pumping until the cylinder is fully pressurised (half way in green).
> 5. Remove cylinder and screw on retaining caps.
> 6. Label cylinder, make note of serial number on gas monitoring proforma and fill in chain of custody form.
> 7. Re-monitor the borehole for gas concentrations and record results.

Note: The steps listed in Example 6.1 are an example.

The Institute of Wastes Management guidance lists the following factors that should be taken into consideration when sampling (Institute of Wastes Management, 1998):

- rate of gas ingress (related to monitored strata)
- water level
- volume of gas contained in the monitoring well – CIRIA publication R131 (Crowhurst *et al*, 1993) advises a minimum of 10 times the volume of monitoring instrument
- instrument's pumping ratio
- length of the sampling probe (which will influence the sampling time and may dilute sample)
- potential depletion of gas in the monitoring point during sampling.

It is standard good practice to monitor the soil gas concentration, pressure and flow rate immediately before sampling and immediately after sampling. By recording these measurements, the ingress of the gas into the installation can be calculated. It will also aid subsequent interpretation of the soil gas regime.

6.2.1 Quality assurance

As with soil sampling there are many factors that need to be considered to provide quality assurance. The main points have been listed in *Quick reference: Additional gas sampling considerations*.

> **QUICK REFERENCE   Additional gas sampling considerations**
>
> Bulk gases are typically present at percentage concentrations and trace gas components at mg/m³ (similar to ppm levels) and ng/m³ concentrations. Hence, quantification of trace components is particularly susceptible to contamination during sample collection, transport and analysis (Environment Agency, 2004b).
>
> To minimise potential contamination, all monitoring equipment must be clean and the equipment must be appropriately designed and constructed (Institute of Wastes Management, 1998).
>
> The samples must be transported in clean, sealed containers with the minimum exposure to other volatile substances before analysis (Environment Agency, 2004b).
>
> Clean monitoring equipment can be achieved by purging the gas sample container before sampling. This ensures that any previous contamination has been removed. If instructed the laboratory can purge the gas sample containers once the sample has been analysed. The use of argon or nitrogen as a purging material is not uncommon.
>
> All samples need to be correctly labelled immediately after sampling to avoid error. A chain of custody form should be filled out and if appropriate the field gas concentrations should be added. This will aid the laboratory in interpretation of the data.
>
> Gas mixtures are not stable, so the samples should be transported promptly to the laboratory.
>
> Before analysis gas samples of both bulk gases and trace components should be stored at ambient temperature (this generally applies to Gresham cylinders and tedlar bags) (Glasgow Scientific Laboratories, 2005).

Once the samples arrive at the laboratory, they should be analysed within a certain time frame. Reactive gases (such as hydrogen sulphide) should be analysed within 24 hours. If these times are not achievable, the laboratory should provide evidence that reliable results can nevertheless be achieved. The guidance varies in regard to time frame within which non-reactive gases should be sampled. These time frames range between three and five days for tedlar bags and up to seven days for Gresham cylinders.

6.3 ANALYTICAL TECHNIQUES TO IDENTIFY THE SOURCE GAS AND HAZARDOUS PROPERTIES

The potential source of gas is typically identified before the ground investigation. In some cases, however, gas may be detected without a known source or with more than one potential source. In such cases sampling and laboratory analysis of the soil gas as detailed above provides data essential to an appropriate level of understanding.

There are many sources of methane in the environment (see Chapter 2). Laboratory analysis of gas samples is essential to identify the site-specific source. By identifying the composition of the gas the potential source/sources can then be interpreted. A summary of the key compounds in each type of methane mixture is shown in Table 8.2 of the Institute of Wastes Management document *The monitoring of landfill gas* (Institute of Wastes Management, 1998).

There are four main analytical methods available for distinguishing landfill gas from methane-rich soil gases:

- gas chromatography
- inductively coupled plasma mass spectrometry
- carbon dating
- stable isotope analysis.

A summary of these analytical methods is provided in Appendix A4, while Figure 6.1, aids the assessor in identifying the most appropriate technique on a site-by-site basis and distinguish between two sets of potential soil gas sources. Generally this figure can be used where hazardous ground gases have been identified on a site but there are potentially other sources present.

Matrix to distinguish between two potential sources

POTENTIAL SOURCE 1	Marsh/peat bogs	Deep peat	Landfill	Made ground	Mines gas	Mains natural gas	Mains coal gas	UG oil/gas reserves
Deep peat	^{14}C							
Landfill	Trace gas	Trace gas ^{14}C						
Made ground	Trace gas	Trace gas	Trace gas					
Mines gas	GC ^{14}C ^{13}C Geology	GC ^{13}C Geology	GC Trace gas Geology ^{13}C ^{14}C	GC Trace gas Geology ^{13}C ^{14}C				
Mains natural gas	GC Pipelines ^{14}C ^{13}C Higher HCs OS	GC Pipelines ^{13}C Higher HCs OS	GC Trace gas ^{13}C ^{14}C Pipelines Higher HCs OS	GC Trace gas ^{13}C ^{14}C Pipelines	GC Trace gas ^{13}C Geology Pipelines			
Mains coal gas	GC Pipelines ^{14}C	GC Pipelines 13C	GC Pipelines 13C ^{14}C Trace gas	GC Trace gas Geology 13C ^{14}C Pipelines	GC Geology Pipelines	GC 13C Pipelines		
UG oil/gas reserves	Higher HCs ^{13}C ^{14}C Geology	Higher HCs ^{13}C Geology	Trace gas ^{13}C ^{14}C Higher HCs Geology	Trace gas Higher HCs ^{13}C ^{14}C Geology	GC ^{13}C Geology	GC Geology Pipelines	GC Geology Pipelines	
UG Fires	GC	GC	GC	GC	GC Geology	GC Pipelines	Pipelines	GC Geology

POTENTIAL SOURCE 2

Key:

UG	Underground
GC	Gas chromatographic analysis of principal gases to determine concentration ratios
Trace gas	GC or GC-MS analysis of trace organic compounds
^{14}C	Determination of $^{14}C:^{12}C$ ratio by mass spectrometry
^{13}C	Determination of $^{13}C:^{12}C$ and 2H:1H ratios by mass spectrometry
Higher HCs	GC analysis of longer chain alkanes
Pipelines	Consult relevant bodies or documentation relating to gas/oil distribution routes
Geology	Consult sources of geological and mining information

Figure 6.1 *The application of investigation methods to gas source identification (Harries et al, 1995)*

The assessor can use the figure to identify the most appropriate technique available to distinguish between these potential sources. As an example, if made ground is present across the site with layers of peat underlying the made ground, the most appropriate technique would be GC or GC-MS analysis of trace organic compounds.

The two most common laboratory analytical techniques are gas chromatography (GC)/gas chromatography mass spectrometry (GC-MS) and inductively coupled plasma mass spectrometry (ICP-MS). The application of these techniques and others are summarised in Appendix A4. In all cases, there are a number of factors that need to be considered before selecting a laboratory or an analytical method. Each of these factors is generally covered by accreditation (ISO 17025:2000). The accreditation ensures:

- documented methods with validation data
- quality control procedures
- trained staff monitored for continuing competence
- equipment files for instruments used in the analysis
- system of internal and external (for example UKAS) audits
- a laboratory quality manual outlining structure and procedures.

(Glasgow Scientific Laboratories, 2005)

6.4 SUMMARY

1. Gas sampling and chemical analyses are relatively inexpensive and can provide important supplementary/validating data to any investigation.
2. There are a number of pressurised and non-pressurised sampling techniques available to enable gas samples to be collected in the field and transported to the laboratory.
3. The adoption of a well-defined sampling protocol will ensure consistency.
4. There are a number of laboratory techniques available to determine the gas concentration measurements and the age/formation measurements.

7 Interpretation of results

7.1 BACKGROUND

There is currently little guidance on the assessment of field measurements and existing guidance provides a variety of approaches:

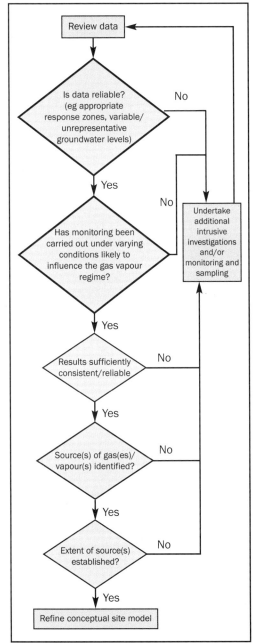

Note: Extracted from Figure 1.1

- the 1991 edition of Approved Document C to the Building Regulations (England and Wales) contained threshold concentrations for methane and carbon dioxide in the ground for which protective measures for buildings were required. Specifically it referred to protection measures for methane concentrations up to 1 % v/v and carbon dioxide concentrations greater than 5 % v/v. Approved Document C referred to, and was supported by, BRE Report 212 (BRE, 1991a) which gave technical guidance on the design and construction of protective measures

- Approved Document C has since been subject to significant revision and a new edition came into force in December 2004 (ODPM, 2004a). The new Approved Document adopts principles of risk assessment for all ground contaminants including methane and carbon dioxide and as such it no longer uses the above threshold concentrations for these two gases. For detailed technical guidance on the design and construction of protective measures the Approved Document refers to the BRE/Environment Agency publication BRE Report 414 (Johnson, 2001) which replaces BRE Report 212 (Department for Communities and Local Government (DCLG), formerly ODPM)

- other documents suggest concentrations within buildings should not exceed the lower explosive limit of the gas

- occupational exposure limits and odour threshold data are also often referred to

- CLEA Soil Guideline Values allow derivation of target values based on toxicological data and site-specific data (for some determinands in soils)

- other CIRIA guidance has suggested use of a "fault tree" to calculate the probability that an event will occur (O'Riordan *et al*, 1995).

The use of defined threshold values may be neither site- nor risk-specific and is contrary to recent guidance CLR11 *Model Procedures for the management of land contamination* (Defra and Environment Agency, 2004a) which promotes risk assessment for ground gases. The application of threshold values has led to an over-conservative assessment of risk that may have resulted in inappropriate remediation. Some recent publications, *Reliability and risk in gas protection design* (Wilson *et al*, 1999) and *Passive venting of soil gases beneath buildings* (Department of the Environment, Transport and the Regions, 1997) has moved towards using a combination of gas concentration and flow rate in assessing risk of soil gases. This has allowed an estimation of the quantity of gas being emitted from the ground. The recent draft guidance from RSK/NHBC (Boyle and Witherington, 2007) introduced a semi-quantitative evaluation system that classifies measured concentrations and flow rates for methane and carbon dioxide into groups of remedial measures. A traffic light colour coding is used to highlight the priority of potential risks and appropriate remedies. The assessment of risk from a valid set of data is discussed in more detail in Chapter 8.

There is little guidance on which or how data obtained from Phase II site investigation should be utilised and assessed in the updating of the initial site conceptual model. This chapter considers this process.

Once a valid set of data has been collected from the field as described in the preceding chapters, it is then necessary to:

- relate point source data to the site as a whole
- place the results into context
- identify the source of the gas
- consider cumulative effects of different sources at a site
- understand how different factors (such as site specific, environmental and meteorological conditions) are affecting the results obtained
- determine the data set to be utilised in the subsequent risk assessment.

7.2 UNDERSTANDING THE COLLECTED DATA

The following sections consider parameters measured as part of a soil gas survey.

7.2.1 Soil gas composition

As described in Chapters 5 and 6, the composition of gas will be measured as part of a soil gas investigation but the means by which it is measured may vary between different investigations depending on the gas being investigated and the investigation objectives. Hand-held monitoring instruments are commonly used which may have differing limits of detection and reporting limits. These should be checked when assessing the results. Also, interference between the gas being measured and other gases can occur. Laboratory analytical methods can be used to determine gas concentrations although they are more commonly used to confirm the results obtained by the hand-held monitoring instruments.

On many brownfield sites where gas emission rates are low, the concentrations of soil gases measured over the period of a monitoring programme can be very variable. Monitoring of such sites will often record concentrations of methane or carbon dioxide, which may vary from zero to occasional peak readings of several per cent. There is currently little guidance as to which reading from the mass of data that has been collected should be used in the risk assessment process (see *Research recommendations* in Chapter 11).

Building Research Establishment (Crowhurst, 1987) have noted that trigger concentrations may be appropriate for contaminants other than gas, as the extent of contamination can be defined and dealt with. It is considered that this is not appropriate with gas contamination because the generation process usually is in a continuous state of flux. It is recommended that maximum concentrations should be used in risk assessments. Chapter 8 discusses which maximum (intra-borehole or inter-borehole maximums) should be used.

The first issue is to understand that the maximum recorded concentration is as close to a theoretical maximum as is possible to obtain, as it is the maximum recorded concentration together with the maximum recorded gas flow that should be used in the risk assessment process; this will represent the worst case scenario. To ensure that the recorded maximum adequately represents the site situation, sufficient monitoring data are required and be measured under different atmospheric conditions. Sladen, Parker and Dorrell have carried out statistical analysis on monitoring data from a brownfield site monitored weekly for three years (Sladen *et al*, 2001). The paper can be used by practitioners assessing small datasets gained from monitoring brownfield sites to compare site-specific mean data with the idealised conditions. This will allow for a check as to whether measured peak readings compare with idealised peak conditions.

7.2.2 Gas flow rate

The purpose of measuring gas flow rates is to predict surface emissions and from these deduce the potential for gas ingress into buildings. The flow rate (measured as litres per hour or metres per second) can refer to either the volume of gas being emitted from a monitoring well per unit time or the rate of movement of gas through permeable strata. A measured borehole flow rate is used to calculate the surface emission rate, but there is little guidance to suggest which of the measured flow rates that have been gathered during the monitoring should be used in the risk assessment.

For instance:

1. Should the concentration of highest recorded gas flow rate at a site be compared with the highest borehole gas concentration even if they occur in different boreholes?
2. Is it more appropriate to use the highest flow rate with the highest gas concentration for each borehole?
3. Are there potentially multiple sources of ground gas on this site which may respond differently to temporal or environmental factors ie should the site be sub-divided into zones to allow interpretation of the ground gas regime?

The assessment of gases is focused on acute effects such as asphyxiation or explosion. As these are "one-off" events it is important to consider the reasonable worst case scenario that may occur as this is when such effects are most likely to be manifested. If the site-wide maximum gas flow rate is not to be used, site-specific factors should be considered to justify why. However, it is also important that proper consideration is given to an assessment of the data (that is to include some "sensitivity analysis"). If such analysis is not incorporated into the assessment and the maxima data are just employed without due consideration, unrealistic results will be generated.

7.2.3 Surface emission rate

Surface emission rate is the volume of gas escaping from a unit area of ground in a unit of time. There two basic approaches in deriving a surface emission rate:

1 Directly measure the surface emission rate with the use of a flux box (further details provided in Chapter 5).

2 Derive the surface emission rate from a mixture of measured data (ie gas concentration and flow rate) and empirical data (assumed zone of influence of standpipe).

7.2.4 Gas pressure

Soil gas under pressure will travel greater distances than gas travelling due to diffusion alone. Increased borehole pressure is included as a risk factor within CIRIA publication R149 (Card, 1996) but there is a lack of guidance on how to interpret the results of soil gas pressure measurements.

Positive pressure readings can confirm gas flow readings. A gas under pressure will migrate greater distances in the sub-surface and be more capable of penetrating gaps in the membranes and migrating via buried structures (Boyle and Witherington, 2007). Negative pressure readings are not so helpful but they can indicate a falling groundwater level.

7.2.5 Gas generation rates

The gas generation rate is the amount of gas produced per unit mass or volume per unit time. An "idealised" gas generation rate can be calculated from laboratory analysis of cellulose and hemi-cellulose content of wastes. It is far more complicated to estimate gas generation rates than gas flow or gas emission rates. The method is also relatively inaccurate and may not relate to the actual gas generation rate for material present within the ground as the actual rate at which gas is generated from soils is subject to a large number of variables such as pH, temperature, moisture content and atmospheric pressure, microbial activity etc. This has led to inaccurate estimates of gas generation potential in the past. Measurements of gas generation rates are uncommon within Phase II site investigation practices.

However, the gas generation rate can inform the risk assessment and reflect gas generation from a landfill or other source. This information can be used to predict the length of time that a risk will be posed to site users. The gas generation rate should be calculated on a site-specific basis where there is a need to further characterise the gas.

In the first instance, the maximum recorded concentration and maximum recordable gas flow rate should be used in the risk assessment process in order to represent worst case scenario. As the understanding of the site develops, it may become appropriate to use other data considered as more representative.

For landfill gas an initial approximation of gas generation can be produced by simply assuming that each tonne of fresh biodegradable waste will produce $10m^3$ of methane per year (UK landfills typically generate 5 to $10m^3/t/yr$). The following calculation will then give an approximation of the rate of methane generation of a landfill. This equation will produce an overestimate of gas flow at peak production and gas flow from historic waste or inert deposits.

$$Q = M \times T \times 10/8760$$

Where:

Q = methane flow in m^3/ hour from fresh waste

M	=	annual quantity of biodegradable waste in tonnes
T	=	time in years during which waste has been placed.
M × T	=	total quantity of biodegradable waste placed over lifetime of landfill.

This is a very simplistic analysis and makes no allowance for reducing rates of generation with time, but it is useful as a first estimation of gas generation to assess the potential scale of a problem. A graph of landfill gas generation (assumed to the combined total of methane and carbon dioxide) for fresh waste was proposed by Nastev (1998) and this also shows an estimation of the declining rate of generation with time (Figure 7.1).

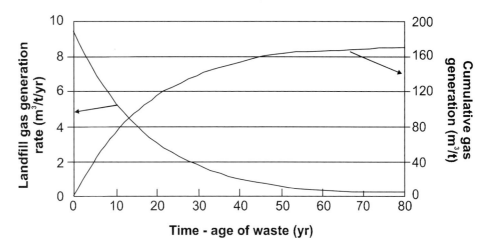

Figure 7.1   *Landfill gas generation against time (Nastev 1998)*

A more accurate estimation of landfill gas generation rates and the changes over time can be developed using the GASSIM computer program developed by the Environment Agency (see Figure 7.2). This uses a first-order kinetic model to estimate gas generation (ie exponential decline), with no lag or rise period, and with waste fractions categorised as being of rapid, medium or slow degradability. This equation (or similar first-order equations) is commonly used in combination with waste input predictions to produce a gas generation profile for the lifetime of the site. Such a multi-phase, first-order decay equation forms the core of the GasSim model (Environment Agency, 2002c).

The equation uses in GasSim is given below:

$$\alpha_t = 1.0846 \cdot A \cdot C_i \cdot k_i \cdot e^{-k_i t}$$

Where:

α_t	=	gas generation rate at time t (m³/yr)
A	=	mass of waste in place (tonnes)
C_i	=	carbon content of waste (kg/tonne)
K_i	=	rate constant (yr⁻¹) (0.185 – fast, 0.100 – medium, 0.030 – slow)
t	=	time since deposit (yr).

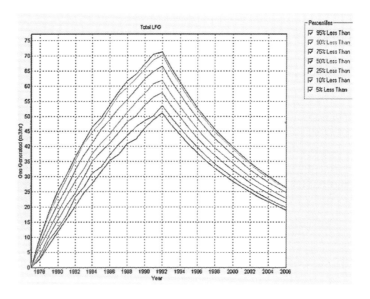

Figure 7.2 *Estimation of landfill gas generation rates with GasSim*

7.3 REFINING THE CONCEPTUAL MODEL

Before starting to estimate risks it is necessary to refine and update the conceptual model of the site with the data from the Phase II site investigation. This will provide a clearer understanding of the pollutant linkages which are realistic and for which risk estimates are needed. An update (following Phase II site investigation) of the initial conceptual model from Chapter 3 is presented in Table 7.1.

The following items should be considered in connection with the nature of the source(s):

1   Which contaminants are present at the site? The borehole monitoring results and bulk gas sample results should be compared to ensure that:

    i.   the data obtained by the portable monitors are sufficiently accurate

    ii. the methane reading displayed by the portable monitor is due to methane and is not due to interference from hydrocarbon vapours and, if so, the hydrocarbon vapours should be identified and quantified.

2   Has groundwater been encountered within the depth of intrusive investigation? If groundwater has been encountered will its vertical movement affect the gas regime? This can occur in a number of ways:

    i.   if the borehole water level is above the slotted part of the standpipe this can lead to false low readings as the soil gas is unable to enter the borehole

    ii. increases in groundwater level or tidal effects on groundwater can also cause increases in gas flow rates and so can increase the potential for migration

3   Is the contaminant recorded associated with a particular source? Review boreholes logs to see if there is an obvious source for the elevated concentrations being recorded (for example natural chalk, peat, volatile odours or the contents of made ground).

4   Is the distribution of the contamination related to any particular source?

5   Nature and characteristics of contaminants – mixtures, volume and concentration present, consistency of results (always the same result in each borehole or generally level readings with occasional peak values).

6   Is there a good correlation between falling atmospheric pressure and higher borehole concentration or flow rate? Typical correlations are:
    - falling atmospheric pressure resulting in gas expansion beneath the ground level, resulting in increased emission rates
    - conversely, rising pressure causing air to flow into the ground, diluting soil gas concentrations.

7   Is there a correlation between rainfall or saturation of soil (ground conditions) and soil gas concentration or flow rate?

In connection with pathways, consideration should be given to the following questions:

1   Is the ground beneath the site (made ground, drift and solid deposits) likely to have significantly high permeability promoting gas flow?
2   At what depth is the gas source present? Is it trapped in a layer with impermeable material above (as can occur with peat)?
3   Does contamination appear to have migrated away from the potential source?
4   Direction of migration of contamination: Is this towards a receptor?
5   Is the contamination migrating along preferential pathways such as services or ducts beneath the site?
6   Are concentrations of contaminants undergoing attenuation along the migration pathway?
7   Are there any potential barriers to migration of contamination between the source and the receptor?

Relating to the potential receptor(s), the following actions are appropriate:

1   Review site plans or development plans to see if there are defined areas on the site where boreholes have similar data (for example all boreholes in the west of the site have methane concentrations over five per cent or all boreholes in the area where flats are to be built have methane concentrations below one per cent).
2   Review surrounding land use to confirm other potential receptors.
3   Consider characteristics and behaviour of receptors (are there small rooms or areas with ignition sources that are poorly ventilated, how much time do occupants spend in different areas, do foundations create pathways?).
4   Confirm which receptors are of priority for the risk assessment.

More information can be found on EA/Defra CLR11 *Model Procedures for the management of ground contamination* (EA, 2004a).

7.4 ADDITIONAL PHASE II SITE INVESTIGATION

As stated in Model Procedures it is essential that the information gathered for use in the risk assessment process meets appropriate quality criteria. The collected data should be reliable in reflecting the true conditions on-site, relevant to the context of the risk assessment, sufficient for the assessment and level of confidence required and transparent. As shown in Figure 1.1 it may be necessary to undertake a further phase of investigation to better understand the soil gas regime in order to carry out the risk estimation process. Whether further data is needed will depend on the views reached when refining the conceptual model and on the approach chosen for risk estimation (see Chapter 8).

The following are examples of additional data which may be required during a further phase of intrusive investigation:

- extent of gas source or contamination – consider installing more boreholes on sensitive property boundaries or in order to clearly define areas of elevated concentrations
- permeability testing of the ground beneath the site – carry out particle size distribution tests and falling head tests if groundwater exists; this will aid understanding of the potential rate of migration in order to provide input to any models used
- information from additional monitoring wells to delineate depth to groundwater and establish presence of any low permeability layers
- further monitoring data to establish concentration trends over time
- additional monitoring techniques to aid in refining the gas regime further (see Section 5.3.2).

Table 7.1   *An example of refined and up-to-date conceptual model summarising environmental risk associated with hazardous gases (see also Table 3.2 for generic conceptual model)*

Source	Pollutant	Receptors	Pathways to receptor	Associated hazard (Severity)	Likelihood of occurrence	Risk
Off-site landfill	Methane, carbon dioxide	Future site users, construction workers	Migration, ingress and accumulation	Effect on human health (Severe)	Unlikely: Landfill 50 m from site boundary. Up to 0.1 % methane and 0.6 % carbon dioxide recorded from boreholes along boundary. Underlying sandy clay likely to inhibit lateral migration.	Moderate/low
		Buildings and structures	Migration, ingress and accumulation	Damage to buildings (Medium)	Unlikely: Landfill 50 m from site boundary. Up to 0.1 % methane and 0.6 % carbon dioxide recorded from boreholes along boundary. Underlying sandy clay likely to inhibit lateral migration.	Low
On-site made ground/ infilled pond	Methane, carbon dioxide	Future site users and construction workers	Ingress and accumulation	Effect on human health (Severe)	Likely: Up to 4 m depth of made ground. Up to 22 % methane and 12 % carbon dioxide recorded. Maximum flow rates 1.8 l/hr. Potential for ingress and accumulation of gas if buildings constructed over area.	High
		Buildings and structures	Ingress and accumulation	Damage to buildings (Medium)	Likely: Up to 4 m depth of made ground. Up to 22 % methane and 12 % carbon dioxide recorded. Maximum flow rates 1.8 l/hr. Potential for ingress and accumulation of gas if buildings constructed over area.	Moderate
		Off-site residents	Migration, ingress and accumulation	Effect on human health (Severe)	Unlikely: Residential properties 20 m distant and underlain by sandy clay. Boreholes outside of infilled pond recorded up to 0.9 % methane and 2.2 % carbon dioxide.	Moderate/low
		Off-site buildings	Migration, ingress and accumulation	Damage to buildings (Medium)	Unlikely: Residential properties 20 m distant and underlain by sandy clay. Boreholes outside of infilled pond recorded up to 0.9 % methane and 2.2 % carbon dioxide.	Low
On-site petrol refuelling area	Hydrocarbons/ volatile organic compounds (VOCs) and semi-volatile organic compounds (SVOCs)	Future site users and construction workers	Inhalation of vapours	Effect on human health (Medium)	Low: Elevated concentrations of flammable gas up to 100 % in localised area close to historical refuelling area. Potential for exposure to construction workers during excavations. Hydrocarbon odour noted during the Phase II site investigation.	Moderate/low (Mitigate by appropriate personal protective equipment (PPE) and site controls)
		Future site users and construction workers	Inhalation of vapours	Effect on human health (Medium)	Likely: Elevated concentrations of flammable gas up to 100 % in localised area close to historical refuelling area. Potential for ingress Hydrocarbon odour noted during the Phase II site investigation. Potential for ingress of vapours if buildings constructed over area.	Moderate
		Future site users and construction workers	Ingress and accumulation	Explosion (Severe)	Low: Elevated concentrations up to 100 % flammable gas during monitoring using infra red analyser in localised area close to historical refuelling area. Potential for exposure during below ground works.	Moderate (Mitigate by appropriate PPE and site controls)
		Future site buildings/ structures	Ingress and accumulation	Explosion (Medium)	Low: Elevated concentrations up to 100 % flammable gas during monitoring using infra red analyser in localised area close to historical refuelling area. Potential for ingress and accumulation of flammable vapours if buildings constructed over area.	Moderate/low
		Ecology	Direct contact, root uptake	Degradation of ecosystem (Medium)	Unlikely: Elevated concentrations detected, localised to historical refuelling area. Present ecosystem well developed and no adverse impacts recorded.	Low
Natural strata alluvial/peat deposits	Methane, carbon dioxide	Future site users, construction workers	Ingress and accumulation	Effect on human health (Severe)	Low: Up to 56 % methane and 4 % carbon dioxide recorded within peat layer, 3.5 mbgl. Overlying sandy clay inhibits vertical migration.	Moderate
		Buildings and structures	Ingress and accumulation	Damage to buildings (Medium)	Low: Up to 56 % methane and 4 % carbon dioxide recorded within peat layer, 3.5 mbgl. Overlying sandy clay inhibits vertical migration.	Moderate/low
Natural strata chalk	Carbon dioxide	Future site users and construction workers	Ingress and accumulation	Effect on human health (Medium)	Unlikely: Chalk present >5 mbgl. Up to 2.5 % carbon dioxide recorded. Overlying sandy clay inhibits vertical migration.	Low

7.5 SUMMARY

Following the preliminary investigation, the monitoring results should be reviewed and the following questions considered:

1. Are the data is reliable? (For example appropriate response zones, presence of high groundwater table).
2. Has monitoring been carried out under varying conditions likely to influence the gas/vapour regime?
3. Are the results consistent/representative?
4. Can the source of the gas/vapour be identified? (It is important to identify the sources but in the case of multiple sources it may not be possible or necessary to accurately attribute the proportional contribution from each source).
5. Can the extent of the source be established? (It is important to identify the approximate extent of the source, but it may not be possible or necessary to accurately define this extent).

If the answer to any of the questions above is "no", further investigation and/or monitoring may be needed.

In the first instance the maximum recordable concentration and the maximum recordable gas flow rate should be used in the risk assessment process, in order to represent the worst case scenario. As understanding of the site develops, it may be more appropriate to use other data which is considered more representative (see Section 7.2).

Similarly practitioners should include appropriate consideration of the uncertainties associated with any calculation in their risk assessments (that is by carrying out appropriate sensitivity analyses) and should also consider obtaining confirmatory data to provide greater confidence in any such calculation.

8 Assessment of risk

8.1 INTRODUCTION

Before embarking upon any qualitative or quantitative assessment of risk the monitoring results should be reviewed and the following questions considered (see also Appendix A5):

1   Are the data is reliable? (For example appropriate response zones, presence of high groundwater table).

2   Has monitoring been carried out under varying conditions likely to influence the gas/vapour regime?

3   Are the results consistent/representative?

4   Can the source of the gas/vapour be identified? (It is important to identify the sources but in the case of multiple sources it may not be possible or necessary to accurately attribute the proportional contribution from each source).

5   Can the extent of the source be established? (It is important to identify the approximate extent of the source, but it may not be possible or necessary to accurately define this extent).

If the answer to any of the questions above is "no", further investigation and/or monitoring may be needed before proceeding with the risk assessment (see Figure 1.1 and extract below).

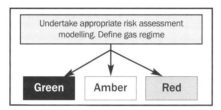

Current risk assessment approaches to soil gas will show the nature of the sources of hazardous gases. Some sources and the associated risk assessments are covered by specific legal requirements from environmental regulators.

These include:

- gas from operational and other licensed landfill sites
- gas from closed and unlicensed landfill sites or from filled sites where made ground is present
- radon
- other natural sources of ground gas
- vapours from VOC contamination.

The risk assessment process for each of these different sources is summarised below. Before proceeding to risk assessment the available data should be reviewed to ensure that it is sufficiently robust and reliable as summarised in the five questions above and as discussed in detail in Chapter 7. If not, then additional data may be required. The risk assessment process may also highlight anomalies in the data and lead to the need for further information.

8.1.1 Operational landfill sites

Landfill gases from operational and proposed waste management facilities (or other hazardous gases that might enter the ground from any industrial activities that fall within the scope of IPPC) are subject to predictive risk assessment from source by the IPPC licence holder. Before a permit to operate can be granted, it should be evident that there would be no adverse impacts on existing developments and preferably no releases from the site. Such risk assessments are usually quantitative, with sound conceptual models, well-characterised source terms and robust assumptions, and sensitivity analyses to demonstrate sufficient factors of safety for all conceivable operating circumstances. However this is beyond the scope of this document and is not discussed further. There are various specific guidance documents on this provided by the Environment Agency in England and Wales and SEPA in Scotland.

8.1.2 Closed landfill sites and made ground

Landfill gases from closed waste sites or where made ground has been placed (where there is no operator and the source can not be reduced), can be controlled by protective measures that seek to intercept gas on its migration pathway towards the development – for example a venting trench. The risk assessment approach discussed in Section 8.2 can be adopted to screen potential risks or classify the ranges of measured gas concentrations and emission rates. This will determine the type of protective measures in buildings, calculate the vent trench dimensions, or determine that development is not suitable for the location.

8.1.3 Radon

Elevated levels of radon are present in the ground in some areas of the UK. However there is comprehensive guidance on this issue and the approach that is to be adopted in terms of risk assessment and provision of protective measures to developments (Building Research Establishment, 1999). This is discussed further in Section 8.6.

8.1.4 Naturally occurring gases

Other gases, such as carbon dioxide, methane and hydrogen sulphide, that are generated naturally underground, are usually unmanageable at their large and diffuse sources. They should be controlled by protective measures within the vulnerable development to prevent public health exposure to unacceptable concentrations or accumulation to potentially explosive levels. The risk assessment process described in this guidance can be used to help the design of gas protection measures where necessary.

8.1.5 VOC vapours

Vapours from volatile land or groundwater contaminants are sometimes dealt with in a similar way to natural soil gases or landfill gas by adopting protective measures in the design of new developments that would exclude all gas ingress. However, for existing developments, this is rarely practical. In any case the favoured remedial solution for managing ground contamination is usually to treat, remove or contain the source rather than implement long-term protection at the point of reception. The assessment then becomes an exercise to test whether or not the soil or groundwater contamination would be of concern to public health at existing concentrations. This is done by evaluating the risk of indoor inhalation exposure based on the characteristics of the contaminant, the type of soil, details of the building construction, and the predicted exposure pattern of occupants.

Detailed quantitative risk assessment is usually required to obtain site-specific assessment criteria and can be rather complex given the transport mechanisms in each of the media along the migration pathway and the large number of variables. Semi-quantitative assessment can be performed as a first step using generic screening values calculated using assumptions about standard soils, building types and critical receptors. Risk assessment for HC vapours is discussed in Section 8.4.

8.2 RISK ASSESSMENT PROCESS

The level of assessment required needs only to be sufficient to support the remedial solution or to justify a decision that no action is required. CLR11 *Model Procedures for the management of land contamination* (Defra and Environment Agency, 2004a) set out the technical framework for tiered risk-based decision-making to accord with government policy on dealing with contamination from historical or natural sources where the opportunity for prevention does not apply. The spectrum of risk management is summarised in Figure 1.1. This helps to show the considerations which may input into a cost benefit analysis to determine the degree of risk acceptability.

When, at an early stage, it is seen that if risk avoidance measures involving remedial work are to be undertaken within the development, the design objectives will often dictate the level and type of assessment. There might be no value in a very detailed assessment to show negligible risk if the execution of such an assessment would outweigh the extra cost of incorporating basic protection measures or evaluating unacceptable risks to a high degree of accuracy if the appropriate solution offers a large factor of safety.

The purpose of risk assessment is to aid judgement and to help make decisions that are:

- legal
- justified
- transparent and understandable.

The reasons for the choice of parameters and other factors in the risk assessment should be clearly stated, preferably in terms that are easily understood by non-specialists. All gas risk assessments and protection designs should be set out in a clear manner so that they can be easily checked (for example by regulators or other consultants acting on behalf of financial institutions).

Site-specific risk assessment process for methane and carbon dioxide is discussed in Section 8.3 and the process for hydrocarbon vapours is in Section 8.4.

To help provide confidence in the risk assessment and gas protection design it is recommended that the proforma provided in Appendix A3 is used to confirm that the collection of the monitoring data cover all necessary parameters, and that this data is comprehensive and appropriately recorded (see also Section 5.3.1).

Gas risk assessment can be complex and practitioners require knowledge of a number of fields including geology, physics and chemistry. In line with guidance produced by the Association of Geotechnical and Geo-environmental Specialists for site investigations and geotechnical assessment, it is recommended that gas risk assessments should be approved by a chartered professional who meets the requirements of a geo-environmental specialist with appropriate experience of contaminated land and gas risk assessments. An equally valid professional qualification is the Specialist in Land Condition (SiLC) (AGS, 1998).

The risk assessment methods described have not been specifically developed to identify contaminated land under Part IIA of the Environmental Protection Act 1990. The risk screening and characterisation process described in Situation A (see Section 8.3.1) could be used to identify sites that are unlikely to pose a significant risk of causing significant harm to a receptor. However the method should not be used to determine sites under Part IIA. The quantitative risk assessment methods discussed in Section 8.3.2 (Situation B) (Step 7) can, and have been, used to help assess the risk posed by sites under the Part IIA legislation, and determine if remediation is required and the scope of that remediation.

8.3 METHANE AND CARBON DIOXIDE

The risk assessment process for methane and carbon dioxide is illustrated in Figure 8.1. For clarity this does not show the site investigation and feedback loops that are integral to the process. It is assumed in this figure that all data is sufficient to support the decision making process.

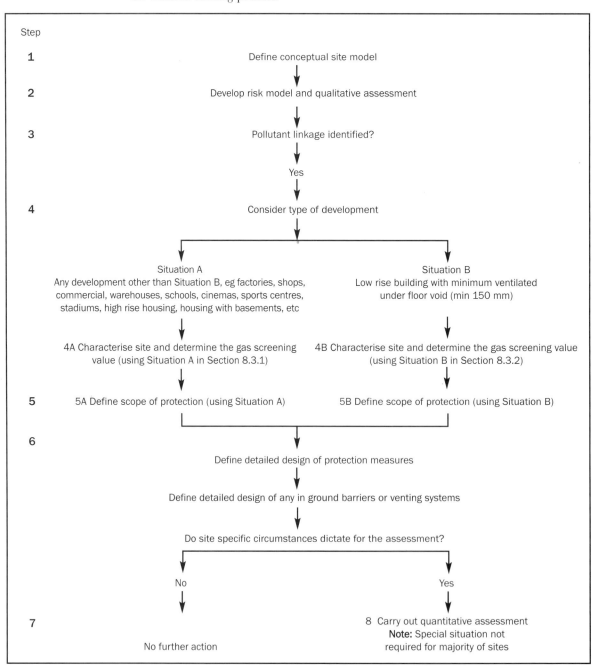

Figure 8.1   *Risk assessment process for methane and carbon dioxide*

Step 1, Defining the conceptual site model

Defining the conceptual site model is discussed in Chapter 3.

Step 2, Risk model and qualitative assessment

A risk model develops the conceptual model to identify potential sources of gas, receptors that could be affected and possible pathways by which the gas could reach the receptors and cause harm.

Once the risk model is developed a qualitative risk assessment can be carried out to evaluate potential risk in descriptive terms. A framework for qualitative risk assessment is provided in CIRIA publication C552 (Rudland *et al*, 2001) and this can be applied to assessment of soil gases. The method requires an assessment of:

- the magnitude of the probability or likelihood of the risk occurring (Table 8.1)
- the magnitude of the potential consequence or severity of the risk occurring (Table 8.2).

Table 8.1 *Classification of probability*

Classification	Definition
High likelihood	There is a pollution linkage and an event that either appears very likely in the short-term and almost inevitable over the long-term, or there is evidence at the receptor of harm or pollution.
Likely	There is a pollution linkage and all the elements are present and in the right place, which means that it is probable that an event will occur. Circumstances are such that an event is not inevitable, but possible in the short-term and likely over the long-term.
Low likelihood	There is a pollution linkage and circumstances are possible under which an event could occur. However, it is by no means certain that even over a longer period such event would take place, and is less likely in the short-term.
Unlikely	There is a pollution linkage but circumstances are such that it is improbable that an event would occur even in the very long-term.

Table 8.2 *Classification of consequence*

Classification	Definition	Examples
Severe	Short-term (acute) risk to human health likely to result in "significant harm" as defined by the Environment Protection Act 1990, Part IIA. Short-term risk of pollution of sensitive water resource. (Note: Water Resources Act contains no scope for considering significance of pollution). Catastrophic damage to buildings/property. A short-term risk to a particular ecosystem, or organisation forming part of such ecosystem (note: the definitions of ecological systems within the draft circular on Contaminated Land, DETR, 2000).	High concentrations of cyanide on the surface of an informal recreation area. Major spillage of contaminants from site into controlled water. Explosion, causing building collapse (can also equate to a short-term human health risk if buildings are occupied).
Medium	Chronic damage to human health ("significant harm" as defined in DETR, 2000). Pollution of sensitive water resources. (Note: Water Resources Act contains no scope for considering significance of pollution). A significant change in a particular ecosystem, or organism forming part of such ecosystem, (note: the definitions of ecological systems within draft circular on Contaminated Land, DETR, 2000).	Concentration of a contaminant from site exceed the generic, or site-specific assessment criteria. Leaching of contaminants from a site to a major or minor aquifer. Death of a species within a designated nature reserve.
Mild	Pollution of non-sensitive water resources. Significant damage to crops, buildings, structures and services ("significant harm" as defined in the draft circular on Contaminated Land, DETR, 2000). Damage to sensitive buildings/structures/services or the environment.	Pollution of non-classified groundwater. Damage to building rendering it unsafe to occupy (for example foundation damage resulting in instability).
Minor	Harm, although not necessarily significant harm, which may result in a financial loss, or expenditure to resolve. Non-permanent health effects to human health (easily prevented by means such as personal protective clothing etc), easily repairable effects of damage to buildings, structures and services.	The presence of contaminants at such concentrations that protective equipment is required during site works. The loss of plants in a landscaping scheme. Discolouration of concrete.

The combination of the two factors is determined using Table 8.3 and the resulting level of risk is described in Table 8.4. The evaluation can be applied to each of the scenarios identified in the risk model and the overall risk assessed. Justification should be provided for all the inputs so that regulators can easily check the model.

Table 8.3 *Combination of consequence with probability*

		Consequence			
		Severe	Medium	Mild	Minor
Probability	High likelihood	Very high risk	High risk	Moderate risk	Moderate/low risk
	Likely	High risk	Moderate risk	Moderate/low risk	Low risk
	Low likelihood	Moderate risk	Moderate/low risk	Low risk	Very low risk
	Unlikely	Moderate/low risk	Low risk	Very low risk	Very low risk

Table 8.4 *Description of risks and likely action required*

Very high risk	There is a high probability that severe harm could arise to a designated receptor from an identified hazard, or there is evidence that severe harm to a designated receptor is currently happening.
	This risk, if realised, is likely to result in a substantial liability.
	Urgent investigation (if not undertaken already) and remediation are likely to be required.
High risk	Harm is likely to arise to a designated receptor from an identified hazard.
	Realisation of the risk is likely to present a substantial liability.
	Urgent investigation (if not undertaken already) is required and remedial works may be necessary in the short-term and are likely over the longer-term.
Moderate risk	It is possible that harm could arise to a designated receptor from an identified hazard. However, it is either relatively unlikely that any such harm would be severe, or if any harm were to occur it is more likely that the harm would be relatively mild.
	Investigation (if not already undertaken) is normally required to clarify the risk and to determine the potential liability. Some remedial works may be required in the long-term.
Low risk	It is possible that harm could arise to a designated receptor from an identified hazard, but it is likely that this harm, if realised, would at worst normally be mild.
Very low risk	There is a low possibility that harm could arise to a receptor. In the event of such harm being realised it is not likely to be severe.

Step 3 Identifying pollutant linkage

Using the risk model the pollutant linkages are identified and a preliminary estimate of risk undertaken. If there is no pollutant linkage identified, then there is no risk. If the estimate of risk for all the linkages and exposure scenarios is very low at this stage then it is likely that no further assessment will be required.

The assessment up to this point can be undertaken using desk study and basic site investigation data such as ground conditions. If further assessment is deemed necessary then gas monitoring in gas wells will be required.

Step 4 Consider the type of development, and Step 5 Define scope of protection

The main method of characterising a site is using the method proposed by Wilson and Card (1999). This is described under Situation A in this guide and can be used for all types of development except for conventional low-rise housing that meets the assumptions made in Appendix A7 (for example it should have a block and beam floor and ventilated underfloor void). For low-rise housing only (Situation B) the method proposed by Boyle and Witherington (2007) should be used. This has been developed for use by the NHBC when classifying gassing sites for development specifically with low rise housing that has a block and beam floor, and a ventilated underfloor void.

As always the assessor should be confident that the gas monitoring results are representative of the likely worst case gas regime on a site and that the data collected from a site is sufficient.

8.3.1 Situation A – All development types except those in Situation B

Step 4A Characterise site and determine the gas screening value

The results of the qualitative analysis can be developed, together with gas monitoring results, to give a semi-quantitative estimate of risk for a site. The system proposed by Wilson and Card was a development of the one described in CIRIA publication R149 (CIRIA, 1996), and is now widely used by consultants and regulators to assess the risks posed by gassing sites. The classification system is summarised in Table 8.5 and is extended to include the general classification of risk obtained from the qualitative assessment. The concept of traffic lights to identify the level of risk, developed by Boyle and Witherington, has also been included.

The method uses both gas concentrations and borehole flow rates to define a characteristic situation for a site based on the limiting borehole gas volume flow for methane and carbon dioxide. In both this guide, and the report by Boyle and Witherington (2007) for the NHBC (2007), the limiting borehole gas volume flow is now renamed as the gas screening value (GSV).

Gas screening value (litres of gas per hour) = max borehole flow rate (l/hr) × max gas concentration (%). For example, monitoring data giving a (maximum) flow rate of 3.5 l/hr and a (maximum) concentration of 4.0 per cent methane would have a GSV of 0.14 l/hr [4.0/100 × 3.5].

The calculation is carried out for both methane and carbon dioxide and the worse case value adopted. The characteristic situation can then be determined from Table 8.5. The higher the classification the greater the risk posed by the presence of gas. It is important to recognise that the GSV is a guideline value and not an absolute threshold. That is, the GSV quoted in Table 8.5 can be exceeded in certain circumstances should the conceptual site model indicate it is safe to do so. Similarly, consideration of the basic factors (such as concentration and flow rate) can lead to consideration of the appropriateness of an increased characteristic situation. Examples of the determination of the characteristic situation for a number of scenarios under Situation A are given in Box 8.1. It is also useful at this stage to compare the results obtained from the risk screening exercise with the results of the qualitative assessment to make sure that there are no anomalies that require further consideration.

An alternative method for defining a gas regime was derived by Owen and Paul in 1998 (see Box 8.2).

Table 8.5 *Modified Wilson and Card classification*

Characteristic situation (CIRIA R149)	Comparable classification in DETR et al (1999)	Risk classification	Gas screening value (GSV) (CH_4 or CO_2) (l/hr)[1] Threshold	Additional factors	Typical source of generation
1	A	Very low risk	<0.07	Typically methane £1 % and/or carbon dioxide £5 %. Otherwise consider increase to Situation 2	Natural soils with low organic content "Typical" made ground
2	B	Low risk	<0.7	Borehole air flow rate not to exceed 70l/hr. Otherwise consider increase to characteristic Situation 3	Natural soil, high peat/organic content. "Typical" made ground
3	C	Moderate risk	<3.5		Old landfill, inert waste, mineworking flooded
4	D	Moderate to high risk	<15	Quantitative risk assessment required to evaluate scope of protective measures.	Mineworking – susceptible to flooding, completed landfill (WMP 26B criteria)
5	E	High risk	<70		Mineworking unflooded inactive with shallow workings near surface
6	F	Very high risk	>70		Recent landfill site

Notes on the use of Table 8.5

1. Gas screening value: (Litres of gas/hour) is calculated by multiplying the maximum gas concentration (%) by the maximum measured borehole flow rate (l/hr) – see Glossary.
2. Site characterisation should be based on gas monitoring of concentrations and borehole flow rates for the minimum periods defined in Table 5.5.
3. Source of gas and generation potential/performance should be identified.
4. Soil gas investigation should be in accordance with guidance provided in Chapters 4 to 6.
5. If there is no detectable flow, use the limit of detection of the instrument.
6. The boundaries between the Partners in Technology classifications do not fit exactly with the boundaries for the CIRIA classification.

Box 8.1 *Situation A examples*

Example 1

Site to be developed for commercial/industrial units with some residential flats and the soil gas investigation has identified a maximum carbon dioxide concentration of 3.0 per cent with a worst-case flow rate of 2.0 l/hr. The gas screening value (GSV) can be calculated as:

- $0.03 \times 2.0 = 0.06$ l/hr

So the site will be characterised as characteristic situation 1.

Example 2

Site to be developed for commercial/industrial units with some residential flats and the soil gas investigation has identified a maximum methane concentration of 4.2 per cent with a worst-case flow rate of 5.0 l/hr. The gas screening value (GSV) can be calculated as:

- $0.042 \times 5.0 = 0.21$ l/hr

So the site will be characterised as characteristic situation 2.

Box 8.1   *Situation A examples (contd)*

> **Example 3**
>
> Site is to be developed for commercial/industrial units with some residential flats and the soil gas investigation has identified a maximum methane concentration of 14 per cent and a worst-case flow rate of 2.5 l/h. The GSV will be calculated as:
>
> - $0.14 \times 2.5 = 0.35$ l/hr
>
> The GSV puts the site in characteristic situation 2. However, the gas concentration is an order of magnitude above one per cent and consideration should be given to whether the site should characterised as characteristic situation 3. The further considerations will typically take into account factors such as the flow rate, the robustness of the data, the source characteristics and the specifics of the development (eg foundation conditions, footprint size) etc. In this case the low flow rate indicates that characteristic situation 2 is an appropriate determination.
>
> **Example 4**
>
> Site is to be developed for commercial/industrial units with some residential flats and the soil gas investigation has identified a maximum methane concentration of 1.2 per cent methane and a worst-case flow rate of 1.5 l/hr. The GSV will be calculated as:
>
> - $0.012 \times 1.5 = 0.018$ l/hr
>
> The GSV puts the site in characteristic situation 1. However, the maximum methane concentration is marginally above the one per cent and consideration should be given to whether the site should be characterised as characteristic situation 2. In this case the further consideration will reflect upon the marginal exceedance of the "typical maximum" value and the very low flow rate. Provided the data was robust (ie the result of a comprehensive monitoring programme) and there was real confidence that the recorded maximum concentration and flow rate was most unlikely to be substantially exceeded, characterisation as characteristic situation 1 would be appropriate.
>
> **Example 5**
>
> Site is to be developed for commercial/industrial units with some residential flats and the soil gas investigation has identified a maximum methane concentration of 69.3 per cent methane and a worst-case flow rate of 1.7 l/hr. The GSV will be calculated as:
>
> - $0.693 \times 1.7 = 1.178$ l/hr
>
> The GSV puts the site in characteristic situation 3. However, the gas concentration is very high and so consideration should be given to whether the site should be characterised as characteristic situation 4. To still progress with a new build at the site, the assessor should be extremely confident that a very thorough site investigation has been carried out. The assessor should also be convinced that the ground gas regime, in particular the flow rates, has been appropriately characterised over a suitable length of time and at the worst-case conditions, and that all data is robust. Further, consideration into all possible methane generation and migration potentials should have been fully characterised within a sound conceptual site model, which should take into account how the ground gas regime (especially flow rates) may be impacted by partial sealing of the site with the buildings and roads of the specific development. In addition, further consideration should be given to appropriate quantitative risk assessment methodologies for the site.

Step 5A Define the scope of protection

The characteristic situation defined in Step 4A can be used to define the general scope of gas protective measures required (Table 8.6). The philosophy behind this is that as the risks posed by the presence of methane and carbon dioxide in the ground increase the degree of redundancy within the type of protective system proposed is also increased, so that if one method or element of protection fails for any reason the building is not exposed to unacceptable risk.

The scope of protective measures proposed by Wilson and Card (1999) is re-assessed in terms of number of protective methods (or levels of redundancy) to allow a less prescriptive approach to detailing protective systems and allow a wider choice in the use of different components. For example a ventilated underfloor void or a positive pressurisation system is one level of protection.

The key issue surrounding gas membranes is their ability to survive the construction process intact (see Chapter 9) and also possibly resist differential settlements. Membranes should be selected based on their performance characteristics and ability to survive the construction phase. An unreinforced 1200 g membrane is unlikely to achieve this and the minimum thickness of gas resistant membrane proposed is unreinforced 2000 g for low-risk sites. The range of protective measures that are available, and their detailed design, is discussed in Chapter 9.

Table 8.6   *Typical scope of gas protective measures*

Characteristic situation (From Table 8.5)	Residential building (not those which belong to Situation B)[1]		Office/commercial/industrial development	
	Number of levels of protection	Typical scope of protective measures	Number of levels of protection	Typical scope of protective measures
1	None	No special precautions	None	No special precautions
2	2	a. Reinforced concrete cast *in situ* floor slab (suspended, non-suspended or raft) with at least 1200 g DPM[2] and underfloor venting. b. Beam and block or pre-cast concrete and 2000 g DPM/reinforced gas membrane and underfloor venting. All joints and penetrations sealed.	1 to 2	a) Reinforced concrete cast *in situ* floor slab (suspended, non-suspended or raft) with at least 1200 g DPM[2]. b) Beam and block or pre cast concrete slab and minimum 2000 g DPM/reinforced gas membrane. c) Possibly underfloor venting or pressurisation in combination with a) and b) depending on use. All joints and penetrations sealed.
3	2	All types of floor slab as above. All joints and penetrations sealed. Proprietary gas resistant membrane and passively ventilated or positively pressurised underfloor sub-space.	1 to 2	All types of floor slab as above. All joints and penetrations sealed. Minimum 2000 g/reinforced gas proof membrane and passively ventilated underfloor sub-space or positively pressurised underfloor sub-space
	3	All types of floor slab as above. All joints and penetrations sealed. Proprietary gas resistant membrane and passively ventilated underfloor subspace or positively pressurised underfloor sub-space, oversite capping or blinding and in-ground venting layer.	2 to 3	All types of floor slab as above. All joints and penetrations sealed. Proprietary gas resistant membrane and passively ventilated or positively pressurised underfloor sub-space with monitoring facility.
	4	Reinforced concrete cast *in situ* floor slab (suspended, non-suspended or raft). All joints and penetrations sealed. Proprietary gas resistant membrane and ventilated or positively pressurised underfloor sub-space, oversite capping and in-ground venting layer and in-ground venting wells or barriers.	3 to 4	Reinforced concrete cast in-situ floor slab (suspended, non-suspended or raft). All joints and penetrations sealed. Proprietary gas resistant membrane and passively ventilated or positively pressurised underfloor sub-space with monitoring facility. In ground venting wells or barriers.
	5	Not suitable unless gas regime is reduced first and quantitative risk assessment carried out to assess design of protection measures in conjunction with foundation design.	4 to 5	Reinforced concrete cast in-situ floor slab (suspended, non-suspended or raft). All joints and penetrations sealed. Proprietary gas resistant membrane and actively ventilated or positively pressurised underfloor sub-space with monitoring facility, with monitoring. In ground venting wells and reduction of gas regime.

Note:

1   For low rise traditional housing with ventilated clear underfloor void (ie Situation B. See Table 8.7, Box 8.4 and Section 8.3.2).

Typical scope of protective measures may be rationalised for specific developments on the basis of quantitative risk assessments.

Note the type of protection is given for illustration purposes only. Information on the detailing and construction of passive protection measures is given in BR414 (Johnson, 2001). Individual site-specific designs should provide the same number of separate protective methods for any given characteristic situation (see Card, 1996).

In all cases there should be minimum penetration of ground slabs by services and minimum number of confined spaces such as cupboards above the ground slab. Any confined spaces should be ventilated.

Foundation design should minimise differential settlement particularly between structural elements and ground-bearing slabs.

Commercial buildings with basement car parks, provided with ventilation in accordance with the Building Regulations, may not require gas protection for characteristic situations 3 and 4.

Floor slabs should provide an acceptable formation on which to lay the gas membrane. If a block beam floor is used it should be well detailed so it has no voids in it that membranes have to span, and all holes for service penetrations should be filled. The minimum density of the blocks should be 600 kg/m^3 and the top surface should have a 4:1 sand cement grout brushed into all joints before placing any membrane (this is also good practice to stabilise the floor and should be carried out regardless of the need for gas membranes).

The gas-resistant membrane can also act as the damp-proof membrane.

Based on Building Regulations Approved Document C (Office of the Deputy Prime Minister, 2004a) which states that *"a membrane below the concrete could be formed with a sheet of polyethylene, which should be at least 300 mu thick (1200 g)"*. Please note the alteration from 300 mm (as stated in the Approved Document C) to 300 mu. 300 mm is a typographical error that has been recognised and corrected for this publication.

The levels of protection referred to Table 8.6 provides a number of protective elements (ie collectively the whole system) that are each capable of protecting a building on their own. In case of failure or damage of one element the remaining element(s) continue to effectively protect the building. The whole system should be designed accurately for the site risk with appropriate interdependence between components and levels of redundancy as determined by a risk-based approach.

The primary method of protection should be the creation of an envelope below the building. Passive ventilation is preferred as it requires less maintenance than any system with fans. This is especially so in freehold/low rise residential developments. Any active system (dilution or positive air) should be able to vent passively in the event of fan failure. The methods of protection are discussed in Chapter 9.

Box 8.2   *Owen and Paul risk assessment methodology*

> A site was derived by Owen and Paul in 1998 as a part of the DETR Partners in Technology *Guide for design research report* (DETR *et al*, 1999). This method is similar to the Wilson and Card approach using both gas concentrations and borehole flow rates to define a gas regime for methane and carbon dioxide. However, by using this approach the calculated regime will be a factor of 10 below the gas screening value derived using the Wilson and Card (1999) approach. This difference is due to the fact that Owen and Paul use the Pecksen approach (1986) in addition to gas concentration and borehole flow rate (Owen *et al*, 1998).
>
> A number of practitioners have raised concerns regarding the assumptions behind the Pecksen methodology, in particular the assumption of a 10 m² zone of influence of a standpipe. There is no doubt that further research is required in this area of the subject. In the meantime, practitioners should make proper consideration of the uncertainties associated with this type of calculation in their risk assessments (that is by carrying out appropriate sensitivity analyses) and should also consider obtaining data on direct surface emission rates to provide greater confidence in any such calculation of gas regimes.

8.3.2 Situation B – Low rise housing with a ventilated underfloor void (min 150mm)

Step 4B Characterise site and determine the gas screening value

The NHBC have developed a characterisation system that is similar to the Wilson and Card system, but is specific to low-rise housing development with a clear ventilated underfloor void (Boyle and Witherington, 2006) (see Table 8.7). This is a risk-based approach that is designed to allow an identification of gas protection for a low-rise housing development by comparing the measured gas emission rates to generic "traffic lights" scenarios. The traffic lights include "typical maximum concentrations" and are provided for initial screening purposes and risk-based gas screening values (GSVs) for consideration in situations where the typical maximum concentrations are exceeded. However, the assessor should carefully evaluate the soil gas regime before proceeding with a design where the typical maximum concentration is exceeded. It should be noted that the method used to develop the GSV thresholds is based on a number of assumptions regarding the proposed structures, and designers should ensure that these assumptions are appropriate to their site. If the proposed low-rise housing development differs significantly from the "model" low-rise housing development (for example, deeper sub-floor void, increased ventilation or larger building footprints), sufficient information should be presented so that the assessor can derive site-specific GSVs. This information is also contained within Appendix A7.

The calculations should be carried out for both methane and carbon dioxide, and the worst case adopted in order to establish the appropriate protection measures. It is also important to note that the GSVs are derived from one air change per day in the sub-floor void providing a simple assessment. As previously stated, if the designer can adequately demonstrate that vent rates are greater (for example, when calculated using BS 5925) then higher site-specific GSVs may be calculated. However, any such alternative assessment should include a sensitivity analysis to take into account the effects of occupiers blocking air vents, for example by construction of patios.

As described in Section 8.3.1 it is important to recognise that the GSV is a guideline value and not an absolute threshold. That is, the GSV quoted in Table 8.7 can be exceeded in certain circumstances (see Footnote 5 to Table 8.7). Examples of the determination of the site characterisation (green, amber or red) under Situation B are given in Box 8.3.

Table 8.7 *NHBC Traffic light system for 150 mm void*

Traffic light	Methane[1]		Carbon dioxide[1]	
	Typical maximum concentration[5] (% v/v)	Gas screening value (GSV)[2,4,6] (litres per hour)	Typical maximum concentration[5] (% v/v)	Gas screening value (GSV)[2,3,4,5] (litres per hour)
Green	1	0.16	5	0.78
Amber 1	5	0.63	10	1.56
Amber 2 / Red	20	1.56	30	3.13

Notes:

1. The worst gas regime identified at the site, either methane or carbon dioxide, recorded from monitoring in the worst temporal conditions, will be the decider for which traffic light and GSV is allocated.

2. Generic GSVs are based on guidance contained within latest revision of Department of the Environment and the Welsh Office (2004 edition) *The Building Regulations: Approved Document C* and used a sub-floor void of 150 mm thickness.

3. The small room eg downstairs toilet with dimensions of 1.50 × 1.50 × 2.50 m, with a soil pipe passing into the sub-floor void.

4. The GSV (in litres per hour) is as defined in Wilson and Card (1999) as the borehole flow rate multiplied by the concentration of the particular gas being considered.

5. The "typical maximum concentrations" can be exceeded in certain circumstances should the conceptual site model indicate it is safe to do so. This is where professional judgement will be required, based on a thorough understanding of the gas-regime identified at the site where monitoring in the worst temporal conditions has occurred.

6. The GSV thresholds should not generally be exceeded without completion of a detailed gas risk assessment taking into account site-specific conditions.

Box 8.3 *Situation B examples*

Example 1

Site to be developed for low-rise housing and the soil gas investigation has identified a maximum carbon dioxide concentration of 3.0 per cent with a worst-case flow rate of 2.0 l/hr. The gas screening value (GSV) can be calculated as:

- $0.03 \times 2.0 = 0.06$ l/hr

The site will be characterised as green.

Example 2

Site to be developed for low-rise housing and the soil gas investigation has identified a maximum methane concentration of 4.2 per cent with a worst-case flow rate of 5.0 l/hr. The gas screening value (GSV) can be calculated as:

- $0.042 \times 5.0 = 0.21$ l/hr

The site will be characterised as amber 1.

Example 3

Site is to be developed for low-rise housing and the ground investigation has identified a maximum methane concentration of 14 per cent, and a worst-case flow rate of 2.5 l/h. The GSV will be calculated as:

- $0.14 \times 2.5 = 0.35$ l/hr

The GSV puts the site in amber 1. However, the gas concentration is three times the typical maximum value for amber 1 and consideration should be given to whether the site should be characterised as amber 2. The further considerations will typically take into account factors such as the value of the gas concentrations and/or the flow rate, the robustness of the data, the source characteristics and the specifics of the development (eg foundation conditions, footprint size), etc.

Example 4

Site is to be developed for low-rise housing and the ground investigation has identified a maximum methane concentration of 1.2 per cent methane and a worst-case flow rate of 1.5 l/hr. The GSV will be calculated as:

- $0.012 \times 1.5 = 0.018$ l/hr

The GSV puts the site in green (by about an order of magnitude). However, the maximum methane concentration is marginally above the one per cent "typical maximum" and, therefore, consideration should be given to whether the site should be characterised as amber 1. In this case the further consideration will reflect upon the marginal exceedance of the "typical maximum" value and the very low flow rate. Provided the data was robust (ie the result of a comprehensive monitoring programme) and there was real confidence that the recorded maximum concentration and flow rate was most unlikely to be substantially exceeded, characterisation as green would be appropriate.

Box 8.3   *Situation B examples (contd)*

Example 5

Site is to be developed for low-rise housing and the ground investigation has identified a maximum methane concentration of 69.3 per cent methane and a worst-case flow rate of 1.7 l/hr. The GSV will be calculated as:

- $0.693 \times 1.7 = 1.178$ l/hr

The GSV puts the site in amber 2. However, the gas concentration is very high, at nearly three and a half times the typical "maximum value" for red. So consideration should be given to whether the site should be characterised as red. To still progress with a new build at the site, the assessor should be extremely confident that a very thorough site investigation has been carried out and that the ground gas regime, in particular the flow rates, has been appropriately characterised over a suitable length of time and at the worst-case conditions, and that all data is robust and beyond scrutiny. Further, consideration into all possible methane generation and migration potentials should have been fully characterised within a sound conceptual site model, which should take into account how the ground gas regime (especially flow rates) may be impacted by partial sealing of the site with the buildings and roads of the specific development.

Step 5B Define scope of protection

Based upon the traffic light classification that is calculated for the site for low-rise housing development only, the scope of protection can be defined using Box 8.4.

Box 8.4   *Gas protection measures for low-rise housing development based upon allocated NHBC Traffic light (Boyle and Witherington, 2007)*

Traffic light classification	Protection measures required
Green	Negligible gas regime identified and gas protection measures are not considered necessary.
Amber 1	Low to intermediate gas regime identified, which requires low-level gas protection measures, comprising a membrane and ventilated sub-floor void to create a permeability contrast to limit the ingress of gas into buildings. Gas protection measures should be as prescribed in BRE Report 414 (Johnson, 2001). Ventilation of the sub-floor void should facilitate a minimum of one complete volume change per 24 hours.
Amber 2	Intermediate to high gas regime identified, which requires high-level gas protection measures, comprising a membrane and ventilated sub-floor void to create a permeability contrast to prevent the ingress of gas into buildings. Gas protection measures should be as prescribed in BRE Report 414 (Johnson, 2001). Membranes should always be fitted by a specialist contractor. As with amber 1, ventilation of the sub-floor void should facilitate a minimum of one complete volume change per 24 hours. Certification that these passive protection measures have been installed correctly should be provided.
Red	High gas regime identified. It is considered that standard residential housing would not normally be acceptable without a further gas risk assessment and/or possible remedial mitigation measures to reduce and/or remove the source of gas.

Step 6 Detailed design of protective measures

Once the general scope of protective measures has been defined detailed design is required. This will include calculation of ventilation rates, fan capacities and required air flow rates. See Chapter 9 for further information. This information will also be required if more detailed quantitative assessment is undertaken (see the following section).

Step 7 Quantitative assessment

Quantitative assessment of risk involves mathematical models and probability analysis to give a numerical estimate of risk. For most sites a risk assessment up to Step 6 will be adequate and provide a cost-effective balance between assessment and proposed scope of protective measures. However, in some cases it may be appropriate to undertake further detailed quantitative assessment if it is likely to reduce the scope of protective measures required or increase the confidence in the proposed protection. It is also a useful technique when considering risks to existing developments from gassing sources.

A statistically valid data set will also be necessary and this means that extensive gas monitoring results are normally required. Quantitative assessment is usually used to justify the decreasing scope of protection, so the minimum periods of monitoring discussed in Chapter 5 will require extending (Figure 8.2). However, the costs of the additional data collection is usually measured in hundreds of pounds whereas the reduction in remediation costs will be measured in tens of thousands of pounds (Figure 8.3). So in many cases it is worthwhile (please also refer to Section 8.2). Further guidance of quantitative risk assessment is provided in Appendix A5.

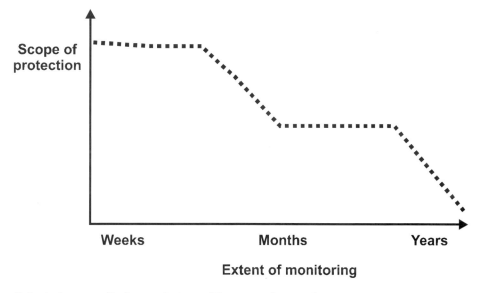

Figure 8.2   *Extent of gas monitoring against possible scope of protection*

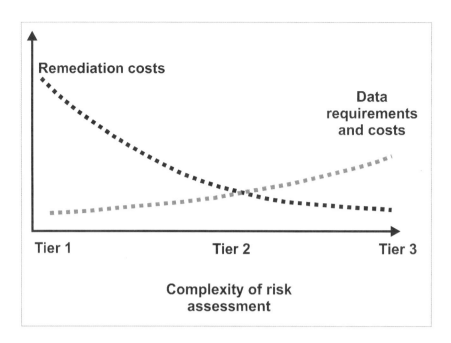

Figure 8.3 *Complexity of risk assessment against remediation costs*

8.4 ASSESSING VAPOURS FROM HYDROCARBON CONTAMINATION

Contamination of the ground by hydrocarbons such as petrol, diesel and coal tars can cause migration of vapours to the surface that pose a risk to health (Figure 8.4).

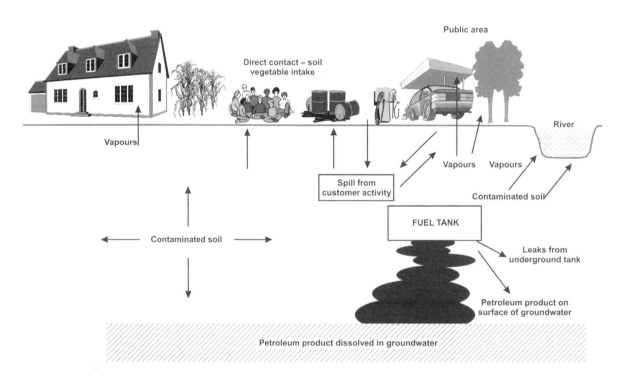

Figure 8.4 *Migration from a leaking underground tank (Institute of Petroleum, 1998)*

In the assessment of potential risks to health from vapours a risk model needs to be developed in the same way as for methane and carbon dioxide. The purpose of the risk model is to characterise the chain of events that complete a pollutant linkage to predict the exposure of a population, group or individual to contaminants through a number of plausible pathways. This results in estimated doses for which the causation of adverse effects is estimated. Attempting to quantify human exposure to any contaminant and the potential adverse effects is highly complex and involves much uncertainty and variability (Defra and the Environment Agency, 2002a). This is particularly true of vapours.

Risk assessment of vapours should follow the available guidance provided by the Environment Agency Contaminated Land Exposure Assessment (CLEA) model (Defra & Environment Agency, 2002a and 2002d). Additional information on risk-based corrective action (RBCA) for vapours is also provided by the American Society for Testing and Materials (ASTM, 2002).

The characterisation of the contamination that causes vapours, based on ground investigation, is usually less precise than defining waste properties and quantities based on landfill records. It should always be remembered that risk assessment is only an estimation procedure to support judgements and draw inferences. No matter how robust the measurement of hazards and consequences might be, risk assessments will always include predictions, expectations, and assumptions based on experience, intuition or statements of belief, that all make the calculation of potential risks fall short of strict objectivity (Hrudey, 1996 and Ferguson *et al*, 1998).

There are multiple exposure pathways between contaminants in soil and groundwater and sensitive receptors, some more tenuous than others. Figure 8.4 illustrates a typical situation where a leaking underground fuel tank can cause harm to various elements of the environment and public health. Figure 8.4 indicates the major routes for contaminants to enter the human body by ingestion, contact or inhalation, which are those included in the Environment Agency Contaminated Land Exposure Assessment (CLEA) model (Defra and the Environment Agency, 2002a, 2002b, 2002c and 2002d). Further guidance on specific aspects of modelling the migration of vapours into buildings and undertaking the risk assessment is provided in Appendix A6.

If a risk assessment is undertaken and it indicates that a site requires gas protective systems to allow safe development, the assessment should be extended to determine the appropriate scope of protection. This is the most commonly neglected part of risk assessment for methane and carbon dioxide, but applies to vapours as well. Particular factors that need to be considered are discussed in further detail in Appendix A6.

8.5 RADON

The simplest qualitative gas risk assessment is the procedure set out in the Building Research Establishment's guidance on radon protective measures for new dwellings (Building Research Establishment, 1999). This method involves checking the location of the proposed development against maps that are based on statistical analysis of indoor air monitoring by NRPB (now part of the Health Protection Agency) and assessment of geological radon-emission potential produced by the BGS. Depending on the colour key of the particular map square (representing a combination of severity and probability, from the NRPB measurements and action levels, and the extrapolation by the BGS to link air data to geology), a decision can be reached on the level of protection necessary in a new dwelling:

- no protection
- basic protection
- full protection.

The classification system is based on a limited number of NRPB measurements and cross-referencing with whole geological units by the maximum radon emission potential in any one 5 km grid square. However, as a commercial service, the BGS offers a "Stage 2" report based on more detailed mapping or a "follow-on" report that gives a site-specific assessment of the radon emission potential of the underlying geology.

The Building Research Establishment's guidance on radon in the workplace (Scivyer *et al*, 1995) relies on building-specific air monitoring data to make the assessment of which protective measures are suitable.

8.6 SUMMARY

1. An appropriately robust and reliable data set is a pre-requisite in any defendable risk assessment.
2. The purpose of risk assessment is to aid decision making. The results of the assessment informs the decision, they do not make the decision for you.
3. A proper understanding of the conceptual site model is critical in any subsequent assessment of risk (qualitative or quantitative).
4. A qualitative assessment of risk (in accordance with CIRIA publication C552) will identify potentially pollutant linkages.
5. Semi-quantitative risk assessment methodologies have been developed which can further elucidate the level of risk and assist the identification of the need for the scope of remedial/protective measures.
6. Both of the methods illustrated (Wilson and Card, 1999, and Boyle and Witheringon, 2007) consider the concentration and flow rate of ground gases. The product of these two factors is termed gas screening value (GSV).
7. The GSVs are guideline values to be used in the assessment of risk. They are not absolute thresholds mandating specific action, professional judgement remains a critical element on the assessment of risk presented by ground gas.
8. There are particular factors to consider in the assessment of risk presented by VOCs and radon.

9 Remedial options

9.1 SETTING REMEDIAL OBJECTIVES

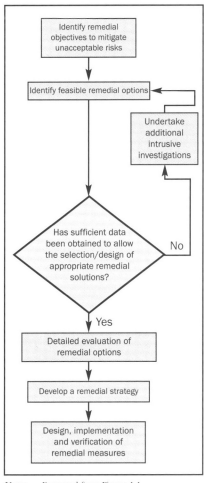

The setting of appropriate remedial objectives is a crucial first step in the design and selection of remedial measures. The objectives will need to reflect the sensitivity of the development (Table 9.1). The gas risk assessment should be used as a starting point for setting such objectives and the subsequent assessment of remedial options which may be applicable to a particular situation. The risk assessment will have identified:

- the nature and extent of gas occurrence (taking into account how well-defined the soil gas regime is and any potential variation outside the measured range)
- maximum and likely gas emissions rates to the proposed development
- maximum and likely gas concentrations
- the existing and future pathways for gas migration
- all existing and future receptors which could be impacted by gas
- construction details of proposed buildings and potential gas entry points.

The remedial objectives need to consider the soil gas regime (and any potential variations), all potential uses of buildings, any likely additions or other changes to building use and the time-frame for which the protection measures are likely to be required. See also Sections 8.3.1 and 8.3.2.

Note: Extracted from Figure 1.1

Table 9.1 *Development sensitivity*

Proposed development	
Houses with private gardens	
Houses without gardens	
Residential building eg flats (not traditional houses)	
Commercial and industrial development	
"Soft" landscaping	
Surfaced hard standing	Increased sensitivity

CIRIA C665

9.2 PHILOSOPHY FOR GAS PROTECTION: BASIC CONCEPTS

Control of gas migration is usually achieved by breaking the migration pathway between the identified gas source and the sensitive receptor(s). The pathway can be broken either at the source or at the receptor. More rarely, the source can be removed and consideration can also be given to any potential for the removal of the receptor.

Control at source typically involves capturing and containing gas emissions within a defined area, and ensuring the gas is safely managed. These techniques are normally applied to management of gas at landfills where, typically, gas extraction wells are installed, and gas is pumped from the wastes and combusted. Such systems are designed to prevent lateral gas migration beyond the landfill boundary and reduce atmospheric emissions of carbon dioxide (Card, 1996).

The second option involves protecting receptors (usually buildings) at risk from gas occurrence. Buildings are usually at surface level, so the requirement for protection is to control vertical migration of gas from the ground. For existing development which is located on gassing land, gas protection measures can also incorporate a programme of monitoring and/or the use of alarms.

The available techniques are illustrated in Figure 9.1 (Card, 1996). Each technique, grouped as source removal, barriers, dilution and dispersion and gas monitoring alarms will be discussed in Sections 9.3 and 9.4 (and associated tables)

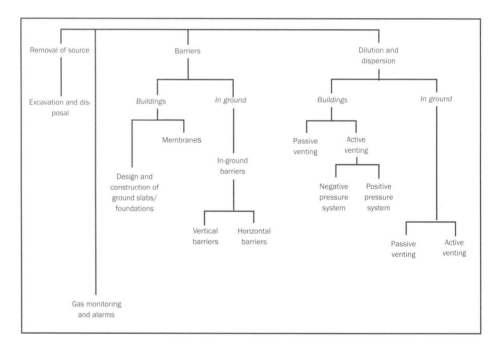

Figure 9.1 *Available techniques for gas protection*

9.2.1 Source removal

In certain circumstances the preferred option for gas control can be to remove the source of hazardous gas from the site. This is likely to be feasible for relatively small, discrete sources of gas-producing material. A significant advantage of such schemes is that, provided any such source is removed in its entirety, there is no potential for failure in the future.

9.3 PASSIVE AND ACTIVE SYSTEMS OF GAS CONTROL (INTERRUPTION OF THE MIGRATION PATHWAY)

Remedial measures that interrupt migration pathways can be classified as either passive or active barriers.

9.3.1 Passive systems

Passive systems rely on creating a permeability contrast between areas requiring gas protection and areas where soil gas can safely vent and dissipate to atmosphere.

Low permeability barriers include:

- naturally occurring clay
- engineered clay
- bentonite enhanced soils
- synthetic membranes, such as LDPE (low density polyethylene) or HDPE (high density polyethylene) (see Figure 9.2)
- engineering/fabricated materials (for example sheet piling).

Barriers can be installed vertically (for example, membrane lining of a vent trench, grout-filled trench, sheet piling) or horizontally (for example floor slab membrane, clay cap).

Figure 9.2   *Installation of HDPE membrane*

There is concern within the industry about the integrity of these gas protection measures, in particular the installation of relatively delicate membranes and the sealing of service entry points. The BRE Report 414 (Johnson, 2001) has identified a number of examples of defective construction including:

- the lack of sufficient sealant where membranes are lapped
- the rupturing of the membrane during installation
- the lack of sealing around service entry points
- the installation of delicate membranes on poorly compacted soils, resulting in pressure points and rupturing as the soils settle.

In addition to identifying typical examples of defective construction, the BR414 report also provides "watchpoints" for each protection method. The "watchpoints" comprise brief instructions, and/or advice, specific to the design and installation of each available gas protection measure.

The concern with poorly constructed membranes and sealing of service entry points has also been demonstrated in Figures 9.3 and 9.4.

Figure 9.3   *Example of poor membrane installation – use of offcuts with insufficient sealing*

Figure 9.4   *Example of poor membrane installation – debris below membrane creating pressure points*

Venting zones can comprise stone or gravel filled pits and trenches, and void space and void formers under buildings. Gas can be vented from safe locations via manholes (at ground level) or vent stacks (often fitted with cowls to increase venting rate of gas from the ground) if elevated dispersion is required, providing a lower resistance pathway for gas emissions to atmosphere. These systems should be properly evaluated (for example using dispersion modelling techniques if appropriate) to ensure that there is no adverse environmental or health effect.

An adsorbent can be inserted in ventilation ducting to reduce emissions of potentially harmful or odorous components of the gas, although by increasing the airflow resistance, this can reduce the effectiveness of passive vents for controlling gas emissions. An adsorbent material will also require maintenance and replacement/ regeneration. However, the use of a high level absorbent/dispersant will also remove odorous compounds.

Gas monitoring linked to alarms could also be considered as a passive system of gas control, but should not be relied upon in isolation to protect developments and is not considered acceptable for low-rise housing.

See Table 9.2 for more details of passive systems.

9.3.2 Active systems

Active systems control gas by either pumping gas from the ground or maintaining a positive pressure of air under or within a building to inhibit ingress of gas. Examples of active systems are:

- perimeter gas extraction wells within a landfill to control lateral migration of gas into surrounding ground
- horizontal gas extraction system to control vertical migration of gas upwards to capping layer and buildings
- installation of positive pressure systems into the ground beneath buildings or into the buildings themselves (for example continuous pressurisation of a cellar to control gas ingress from surrounding ground).

See Table 9.3 for more details of active systems.

The Building Regulations state that actively ventilated systems are generally not appropriate for private housing (Office of the Deputy Prime Minister, 2004a).

9.4 DETAILS OF SYSTEMS AVAILABLE

Many gas protection measures are well-established and easy to incorporate within the overall design of a new development. The appropriateness of a particular protection system relates to the type and sensitivity of the proposed development and the gas regime defined within the ground. A summary of the key features of each system, along with details of further references, is presented in Tables 9.2 and 9.3.

9.4.1 Passive systems

The general characteristics of passive systems are that they:

- utilise the concept of a permeability contrast with venting to allow dissipation of gas to atmosphere in a managed, safe way, and a barrier system preventing gas migration to identified receptors
- employ a multi-barrier concept to include levels of redundancy in the protection system, in case of failure of one component
- are a single installation which should be quality assured and verified by third party
- are low maintenance
- avoid the need for any back-up systems

- are suitable for residential development and other developments where institutional control of management system is likely to be difficult
- are used to protect receptors where source term control is difficult or not possible (for example in areas affected by gas from coal workings)
- can be supplemented with absorbent for control of hazardous/odorous materials.
- are generally not suitable for use under conditions of high gas emission rates from the ground or large and/or complex sub-structures

Descriptions of passive systems are summarised in Table 9.2.

Table 9.2   *Summary of passive systems*

Barrier type	Typical properties and features	Typical applications and use	References for further design details
Natural clay	Natural clay is an effective low permeability barrier to gas movement. Moist clays have typical permeabilities of less than 1×10^{-9} m/sec. Vertical and horizontal barriers available. The major problem with clay is the variations in pore size and pore distribution. In addition clays can potentially crack and dry out creating migration pathways for the soil gas.	Natural clay is utilised as a barrier against lateral migration of soil gases. Many landfills are situated in worked out clay pits. Natural clay strata can be used as a low permeability "buffer zone" between a known source of soil gas and a proposed development. Clays overlying organic silts and peats, which are gassing, and coal measures areas, can be utilised as a natural barrier. However, all clays do have a finite permeability and will transmit gas (albeit at very low flow rates) depending on the clay type.	Further details can be found in: - Card, G B, 1996 - Barry, D L, Summersgill, I M, Gregory, R G and Hellawell, E, 2001 - DETR *et al*, 1997 - CIRIA, 1996b - Building Research Establishment, 2004 - Johnson, 2001.
Reworked/ engineered clay	Clay can be an effective low permeability barrier when it is reworked, relaid and compacted to form a barrier layer. This effectiveness can be tested /verified by on site measurement of density and laboratory testing (for example permeability). Homogeneity can be controlled in this process to provide a barrier of uniform thickness and composition. Engineered clay layers can incorporate a plastic membrane "sandwich", which will reduce the risk of the deeper clay drying out and cracking.	Clay caps are often used across restored landfills, and on development sites as a barrier to control exposure to sub-surface contaminants. Such barriers will also be effective in controlling gas migration, although prevention of vertical migration will tend to push soil gas sideways, which could increase the risk of gas migration beyond the boundary of the site or development area.	Further details can be found in: - Card, G B, 1996 - Department of Environment, 1989 - Environment Agency, 2004 DETR *et al*, 1997 - Building Research Establishment - CIRIA, 1996a,2004 - CIRIA, 1996b.
Synthetic membranes	Synthetic membranes comprise PVC, uPVC, LDPE, HDPE and composite liners reinforced to minimise elongation with aluminium layers to reduce permeability. Membrane liners are installed to prevent the ingress of gas through the building fabric, through features such as shrinkage cracks, construction joints and porous construction materials. Independent validation of these materials should be considered.	Established use in floor slabs of new buildings. Proven for housing and industrial/commercial developments. Care should be taken during the installation of membranes due to the potential for damage to occur.	Further details can be found in: - Card, G B, 1996 - Johnson, 2001 - CIRIA, 1996a.
Grout injection and slurry walls	Injection of low permeability grout into soils, or backfill of trenches.	Typically used to prevent lateral migration of gases either off site from a raised area, or to prevent migration onto a site from an off-site source.	Further details can be found in: - Card, G B, 1996 - CIRIA, 1996a.
Open Voids	Free venting floor space under buildings to create permeability contrast. Normally used in conjunction with membranes.	Modelling shows that an open void has the best venting characteristic of any passive below building venting layer. Floor voids under houses are usually vented via air bricks. They depend on barriers created by/within the floor slab of the building.	Further details can be found in: - Card, G B, 1996 - Barry, D L, Summersgill, I M, Gregory, R G and Hellawell, E, 2001 - Johnson, 2001 - DETR, 1997.

Table 9.2   *Summary of passive systems (contd)*

Barrier type	Typical properties and features	Typical applications and use	References for further design details
Granular venting layers	Gravel venting layers, sometimes fitted with perforated pipes to aid in collecting and removing gas from building.	Used below buildings for gas dispersal, within sites as high permeability venting zones to dissipate gas to air, and in venting trenches to intercept lateral migration of soil gas. Vent trenches can be fitted with a membrane barrier along one face to reduce the risk of onward gas migration beyond the trench. There is some evidence that, over time, gravel venting areas will encourage growth of bacteria which microbiologically oxidise methane to carbon dioxide within the gravel matrix. Bacteria could also be introduced via soil particles being washed in the gravel voids.	Further details can be found in: • Card, G B, 1996 • Johnson, 2001 • DETR, 1997.
Synthetic void formers	Formed of polystyrene boards and capsulated polyethylene/ polypropylene geosynthetics, to provide structural support combined with venting characteristics. Permeability is significantly higher than gravel venting layers	Under buildings as combined load bearing and gas dispersion medium. Within trenches as an alternative to gravel backfill. This reduces the volume of excavated material requiring re-deposit/ disposal and avoids the use of gravel resources.	Further details can be found in: • Arup, 1999 • BRE, 1995a • BRE 1995b.
Gas dispersion via manholes or vents in safe locations	Manholes and vents can be provided to relieve gas pressure. Ground-level orifices need to be provided at locations where members of the public will not be present. Vertical vents can be concealed within structures such as lamp-posts or sign-posts. Installing an adsorbent material such as activated carbon in the ductwork can be effective in reducing emissions of volatile organic substances which may be giving rise to adverse odours or is potentially associated with adverse health effects. The adsorbant will need maintenance and regeneration to retain its effectiveness.	As a supplementary control feature, or to deal with situations with existing odour problems.	Further details can be found in: • Card, G B, 1996.

9.4.2 Active systems

The general characteristics of active systems for gas control are that they:

- provide for the continuous extraction of gas from the ground, or the maintenance of a positive pressure curtain within or beneath buildings and confined areas
- can include the combustion of collected methane and subsequent dispersion of combustion products
- require management and disposal of condensate which dispenses from the collection pipework.

Active systems (see Figure 9.6) of gas management require the following:

- clearly-defined responsibility and programme for on-going management, control and maintenance of the gas management system for life of building
- a source of power to operate pumps and fans
- rapid, contingency response to a power failure (back up power supply sometimes required) including designated evacuation procedure should complete failure occur
- a programme of on-going management and maintenance
- provision for replacement of system components with a finite life expectancy (for example, replacement of gas wells)
- security to prevent interference and vandalism
- flexibility to respond to changing soil gas regimes.

For these reasons, active systems require institutional control and are generally not appropriate for long-term solutions for residential development. Where they are to be used, specialist design input should be obtained. Features of typical active systems are summarised in Table 9.3.

Figure 9.6
Installation of positive pressurisation pipework system and completed control panel

Table 9.3   *Summary of active systems*

System type	Properties and features	Typical applications	References for more information
Positive pressurisation system below buildings	These systems rely on the provision and maintenance of a positively pressurised zone of clean fresh air under the building maintained by a small, low power fan or pump which runs constantly and replaces loss of pressure	Particularly appropriate on large industrial and commercial buildings or where complex sub-structures make provision of positive cross-ventilation difficult or impossible.	Further details can be found in: • Stevens and Crowcroft, 1995 • Johnson, R, 2001.
Vertical gas extraction systems	Systems are used to control lateral migration of gas in soils, usually associated with landfill sites. Typical systems comprise a series of vertically drilled gas collection wells fitted with well-heads, linked to gas collection pipework, pump and flarestack. Gas wells comprise vertically drilled boreholes fitted with perforated collection pipe surrounded by a gravel pack. The well-head assembly allows control of gas flow from the well. Water vapour from extracted gas condenses in collection pipework. The collection system needs to include condensation traps and removal points. Utilisation can be for direct use or power generation. Best practice guidance from the Government requires burning of methane (within utilisation systems or flarestacks) to convert methane to carbon dioxide and reduce its global warming potential.	Typically installed in landfill sites, where the rate of gas generation and volume produced are high. Such sites are unlikely to be considered suitable for redevelopment with buildings until gas generation rates have fallen to levels where active extraction of gas from the ground is no longer required. EA and SEPA guidance specifies active collection and extraction of landfill gas will be required where total site gas emissions are higher than 50 to 100 m^3/hour.	Further details can be found in: • Card, G B, 1996 • Department of Environment, 1989.

Table 9.3   *Summary of active systems (contd)*

System type	Properties and features	Typical applications	References for more information
Horizontal gas collection systems	This is a similar concept to vertical active systems, except that gas is collected in perforated pipework laid horizontally near the surface of the land. The objective is to control vertical, rather than lateral migration of gas and protect receptors developed on the surface of the site. Gas is pumped at a much lower rate than for vertical systems to minimise the risk of air ingress. Condensate can build up in the pipework.	This technique is used much less frequently than vertical collection systems. However, one example of its use is at the motorway service station at Thurrock, developed on an old landfill site, where a horizontal gas collection system was installed below the capping layer. Horizontal systems can be installed to collect fugitive gas emissions to atmosphere, and have been used to capture gas emissions giving rise to odour nuisance.	Further details can be found in: • Card, G B, 1996 • Department of Environment, 1989.
Forced ventilation below buildings	These systems comprise fans which blow fresh air through a void beneath a building. Fans can operate continuously, or can be linked to a gas sensor and activated at a pre-set gas trigger level (such as 20% LEL).	There are a few examples where this system has been used as the primary gas protection measures (motorway service station at Thurrock) incorporating a continuously ventilated under floor void space for the amenity building, These systems are sometimes used in emergency situations where gas is detected beneath existing buildings. Past systems have pumped exhaust gas from combustion of methane under buildings to provide a positive pressure curtain to prevent emissions of methane from underlying soils. This is not advocated, since combustion products including carbon dioxide and carbon monoxide are introduced under pressure below buildings.	Further details can be found in: • Card, G B, 1996.
Interior positive pressure systems	These work on the principle of creating positive pressure inside a building or confined area, such that the potential for gas flow is always outwards and away from the building or confined area. Continuous warm air heating or air conditioning will achieve this effect. In addition, specialist units are available. A drawback of this technique is that opening of doors or windows can equilibrate the internal and external pressure, negating the effects of the positive air flow.	Positive pressure systems have been installed in houses to control gas ingress, normally in basements and cellars	Further details can be found in: • Card, G B, 1996 • BRE, 1995c. Limited information can be found in: • Barry, D L, Summersgill, I M, Gregory, R G and Hellawell, E, 2001.

9.5   MONITORING AND ALARMS IN BUILDINGS

The installation of permanent monitoring instrumentation in buildings (both residential and commercial), sometimes linked to an alarm system, is normally utilised only in circumstances where existing buildings are located on gassing land. In these cases, monitoring and alarm systems are retro-fitted and are utilised to support the primary gas control measures especially where there is doubt that the current control measures are sufficient. The objective of the monitoring/alarm system is to ensure that the soil gases do not accumulate to potentially hazardous concentrations in the protected building.

It is essential to be aware that gas alarms are installed to warn of possible accumulation of hazardous gases in buildings and confined areas. Alarms do not prevent gas entry, and should not be considered as a means of gas control.

The design of monitoring/alarm systems should take into account:

- the instrumentation type
- the location of the detector
- the longevity of the system
- its calibration, maintenance and security in the long-term etc.

Typically methane alarm systems incorporate two trigger levels:

1 **Low level.** 0.25 – 1 % v/v (5 – 20 % LEL) and related to switching on active systems and/or call-out of service personnel
2 **High level.** 1 – 2.5 % v/v (20 – 50 % LEL) requiring immediate action (such as evacuation of the premises).

Useful guidance on these systems is provided in CIRIA publication R149 (Card, 1996) and BRE guidance (Johnson, 2001).

The use of alarm systems can give rise to a number of potentially difficult issues which need to be taken into account. In recent years such monitoring/alarm systems have normally been installed only *in extremis* on existing sensitive developments. It is highly unlikely that serious consideration would be given to proposals for development which incorporated an in-building monitoring and alarm system as part of an active planned gas protection measure. This is principally due to the following issues:

- an increase in the perceived level of risk due to the presence of the alarm and/or an evacuation procedure
- the potential for erroneous/false alarms
- the potential for accidental/deliberate interference with the alarm system (for example switching it off)
- the need for continued calibration/maintenance (and the associated intrusion).

9.6 SUMMARY

1 Setting appropriate remedial objectives is crucial. These objectives should consider the gas regime, the certainty of its definition/potential for variation, the potential uses of the buildings/land and the time frame.
2 The control of gas migration can be achieved by source removal or by breaking the migration pathway at the source or at the receptor.
3 Both passive and active systems are available to control gases.
4 Active systems are not normally suitable for privately-owned residential developments.
5 Alarms do not prevent gas entry and should not be considered as a means of gas control. In existing buildings permanent monitoring and alarm systems are normally considered as a support to positive measures to control gas.

10 Post development monitoring

10.1 CONSTRUCTION MONITORING

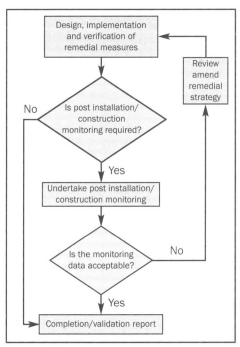

Note: Extracted from Figure 1.1

Post development monitoring is described in CLR11 *Model Procedures* (Defra and Environment Agency, 2004a) as the process to investigate the effectiveness of remediation, to confirm predicted behaviour as an early warning of adverse trends and to maintain remediation to ensure continued functioning and effectiveness in accordance with the original design (Defra and Environment Agency, 2004a). With respect to soil gas, particular consideration needs to be given to a variety of issues and circumstances when assessing such monitoring programmes.

The risk assessment process will have identified the time for which gas protection is likely to be required at a particular development. The protective measures should be designed for the appropriate lifetime of the development if there is uncertainty in the time over which gas and vapours will occur at the site. Although the Model Procedures recommend post development monitoring it is recognised that particular circumstances may prevail such that monitoring and/or maintenance is not applicable (Defra and Environment Agency, 2004a). The Model Procedures include no reference to soil gas in this respect. However, the temporal variability associated with hazardous soil gases, and the nature of the hazard (that is the possibility of the hazard relates to the potential for realisation within the dwelling itself), can make post-construction monitoring of a new development site both undesirable and impracticable once the construction of residential dwellings is completed (and the homes are ready for sale and occupation). This is discussed in more detail in Sections 10.5 and 10.6.

For existing developments, where potential hazards have already been identified and remedial measures have been retro-fitted, post-construction monitoring is more readily incorporated into a long-term management plan.

The following text describes the guidance that currently exists with respect to monitoring. It then outlines how the effectiveness of gas remediation measures may be affected by time or by changes which could take place both on- and off-site. Comment is then made on the impact that the public perception of risk may have on a monitoring strategy. The chapter concludes by making recommendations about the need for and scope of long-term monitoring of developed land affected by gas.

10.2 EXISTING GUIDANCE

There is little specific guidance on post-construction monitoring. UK Waste Management Paper 27 (WMP 27) (Department of Environment, 1991), now superseded by more recent EA guidance (Environment Agency, 2004a), specified criteria for landfill stabilisation. This states that stabilisation could be considered to have been achieved if monitoring on four occasions, over a period of more than 12 months, revealed methane concentrations to be maintained below 1 % v/v and carbon dioxide concentrations below 1.5 % v/v.

The Environment Agency have produced a consultation document on landfill completion (Environment Agency, 2003a). The document includes details of waste testing to establish completion criteria, advocates monitoring for soil gas composition and surface emissions, and relating results to atmospheric pressure, sensitive receptors and "natural" baseline values of gas. The document acknowledges the difficulties in gaining sufficient statistically representative data from waste analyses. The approach is risk-based and the document does not set specified values for methane, carbon dioxide or other parameters which, if met, are indicative of landfill stabilisation.

The Building Regulations do not include any specific information for post-construction monitoring of protective measures. It does state that actively ventilated systems are generally not appropriate for private housing (Office of the Deputy Prime Minister, 2004a).

The BRE/Environment Agency report, *Protective measures for housing on gas-contaminated land* (Johnson, 2001), indicates that there are usually no maintenance or control procedures for gas protective measures within a residential development. However, BRE do recognise that the effectiveness of gas controls can be compromised by actions of the householder (intentional and inadvertent).

CIRIA publication R149 *Protecting development from methane* (Card, 1996), addresses the requirements for long-term management of gas control systems. The long-term management process is defined as "the process through which a gas-control system should be operated in order to ensure that":

- required performance standards are identified
- appropriate maintenance and servicing of gas-control systems are identified and planned in order to continue to meet performance standards
- appropriate remedial action is undertaken if performance standards are not achieved.

The report highlights the need for long-term management of the chosen gas control system and recommends the minimum scope for the operations manual detailing the gas control management. However there is little specific information on the need for, and scope of, post-construction monitoring other than that the design monitoring programme should be predicated on a risk-assessed basis.

10.3 FUTURE CHANGES TO THE DEVELOPMENT AND IMPACT ON SOIL GAS REGIME

The risk assessment should include, as far as is practicable, known and potential changes in site use which could affect the gas regime or compromise the long-term effectiveness of gas protection and control measures. These potential influences can then be incorporated into the final design of remedial measures and any associated requirements for verification and/or post construction monitoring. However, it is accepted that all possible events and future developments cannot be predicted and planned for in the design of protective measures. Any such future changes would normally require re-assessment of risk and the need for and scope of remediation measures.

The following influences could apply:

1 Effects on existing gas protection measures.

2   Effects resulting from changes in site use.

3   Effects resulting from changes off-site.

These influences are discussed below.

10.3.1 Effects on existing gas protection measures

The potential effects and/or influences that could apply to existing gas protection measures are listed below:

- clogging of granular and synthetic venting areas with sediment and biomass, and clogging of gas extraction wells, which will reduce venting efficiency. *This can be predicted, and the maintenance schedule should allow for regular inspection and rehabilitation/replacement of venting media*

- rupture of membranes as a result of ground settlement or perforation during building works. *Any potential for settlement should be established at design phase, since it will need to be addressed as part of the foundation design of the development. In areas, where settlement could occur (for example resulting from shrinkage of clays), membrane design should accommodate some settlement. Damage of the membrane by individual householders is less easy to prevent. A description of the protective measures and need to avoid damage to the membrane can be included as part of the sale particulars of individual properties*

- increase in water table or flooding which will reduce efficiency of venting areas. *Developments in areas of known flood risk should be required to be protected from flooding as part of the planning procedure. The gas control measures should accommodate additional safety factors to allow for potential increased gas production resulting from flooding impacts*

- blocking of sub-floor ventilation panels and air bricks can create a risk of gas build-up within the buildings, if the air bricks are the only form of gas protection measures. The blocking of air bricks can be caused by vegetation growth and/or placement of materials alongside walls. *Information and advice can be included in sale particulars of the property*

- wear and tear and need to replace moving parts in active systems (for example fans). *This should be foreseen at the design stage, and incorporated into the maintenance schedule.*

10.3.2 Effects resulting from changes in "site use" or "extending the building footprint"

Changes that occur post original construction can affect and/or influence the gas regime:

- new extensions and building work (eg conservatories, house extensions, outbuildings) can block existing venting and may not be fitted with comparable standards of gas protection. *Construction may be controlled through the planning process (provided the development is of the scale to require planning permission). Information with sale particulars will include risk impacts of such developments, and information on how to obtain appropriate advice on acceptable design. However, there remains a potential for building work to be carried out (for example the erection of garden sheds) that does not require planning permission)*

- covering of open areas with hardstanding (eg for car parking) can reduce natural ventilation capacity, and encourage lateral migration of soil gas/vapours. *The planning authority should hold records of the existing development, and gas protection provided. Hardstanding can be designed to be gas permeable, or additional venting can be incorporated around the hardstanding area.*

Note that there are sometimes conflicting development demands, for example impermeable surfaces to reduce infiltration of rainwater to protect groundwater resources but, on the same site, the need to allow venting of soil gases. Both objectives need to be satisfied without comprising the functionality of the other

- use of soakaways can increase water infiltration into soils with potential to increase the rate of gas production by increasing moisture content of buried organic material (natural and anthropogenic). *Drainage requirements should be controlled through the planning process.*

10.3.3 Effects resulting from changes off-site

Changes in land use surrounding the site can also affect the soil gas regime. These effects and/or influences are listed below:

- redevelopment or change in use of adjacent land could result in an increase in gas migration to the development site if surface venting is restricted. *Awareness and development restrictions should be held by the planning authority*

- conversely, incorporation of gas control measures on adjacent land may improve venting capacity of the area, resulting in a reduced risk to the development. *Awareness and development constraints should be held by the planning authority.*

Impacts can be reduced, where feasible, by incorporating planning conditions into further site development which require gas protective measures to be installed in new extensions and building work. Developers should ensure the local authority (planning department and building control) is fully aware of the type and level of gas protective measures installed at developments, such that new applications can be assessed. In some circumstances, the local authority may require regular monitoring of the soil gas regime and regular assessment of the continuing effectiveness of building protective measures. The authority may set its own risk criteria, below which monitoring can be discontinued, or the need for future gas protection to be re-assessed.

10.4 RISK PERCEPTION ISSUES

The perception of risk is affected by a large number of factors, both technical and non-technical (further details in Rudland *et al*, 2001 and Ferguson *et al*, 2003). The effectiveness of communication is particularly important in the stakeholder's opinion of any stated level of risk (and the risk assessment process itself). In addition, the perceived level of risk will reflect the development status of the site. That is, the risk perceived by a resident who has lived on a site for a number of years where a hazardous gas regime has just been discovered may differ greatly from that of a person considering a move to a site where gas protective measures have been designed and installed into a new development.

Issues of perception also need to be taken into account when determining the need for and scope of any long-term monitoring regimes. For example, the presence of a monitoring programme can be seen to imply that the hazard has not been managed by the remedial design/works and that potentially significant risks remain. The conclusion of a monitoring programme within an appropriately well-defined period can be seen to deliver "closure". Some examples of the potential issues that may be considered by the various parties are summarised in Table 10.1.

Table 10.1   *Summary of potential concerns*

Stakeholder group	Potential concerns
Owner occupiers and landlords	Personal safety Maintenance of property value Ability to sell property
Tenants	Personal safety
Developer	Maintenance of reputation of company/individual developer Resource and cost impacts of on-going monitoring
Development funders	Maintaining value and ability to sell
Regulators	Safety of occupants Resource and cost impacts of ongoing monitoring
Insurance companies	Limitation of claims arising from gas related incidents

10.5 PRACTICAL ASPECTS OF THE LONG-TERM VERIFICATION MONITORING

The "long-term" or indefinite post installation monitoring of the effectiveness of gas protective measures may be considered neither desirable (the occupier never achieves closure) or necessary from a risk mitigation/cost benefit viewpoint. Furthermore, routine monitoring of occupied premises, both residential and commercial (internal to the premises, in any sub-floor void space, or in private gardens etc), would be considered impractical and/or unduly intrusive in all but extreme circumstances.

The ideal objective of post-development/post-remediation monitoring may be to provide data confirming that the installed protective measures are effectively excluding hazardous gas to the design criteria. However, due to the practical aspects discussed briefly above, realisation of such an objective is unlikely to be practicable once a dwelling has been occupied (or a new home is ready for sale or occupation). So it is considered more practicable for a twin track approach to be adopted.

The first element of this relates to the assessment of the soil gas regime and the second element to its long-term performance/variability.

1. It is considered essential that the potential variabilities of the soil gas regime are properly recognised and taken into account in the design of gas protective measures, discussed in Chapter 9.
2. The objective of the post-construction monitoring can be considered as confirmation that the soil gas regime has remained within the parameters for which the protective measures have been designed (and there is no evidence that this is likely to change for example rising trend etc). A programme to deliver such an objective is outlined in Section 10.6.

10.6 RECOMMENDATIONS FOR POST-CONSTRUCTION/POST-REMEDIATION MONITORING

It is recommended that a programme of post-construction soil gas monitoring is carried out and reported in the completion report for both development sites and sites on which remedial works have been implemented. The objective of the monitoring programme is to confirm that the soil gas regime remains within the design parameters of the protective measures. Such a programme is not envisaged to normally extend beyond the point of sale/occupation of any development. Typically such "post-construction" monitoring would take place during and after the earthworks/foundation construction and during the period over which the above ground building is constructed. At some point before the sale (and before occupation) a completion or verification report (see Section 10.7) would be required.

For example, the NHBC (when carrying out the building control function) are required to comply with the Council of Mortgage Lenders' requirements which, in essence, mean that the property should be "completed" before occupation. A completed property is one which does not have outstanding items that would:

- present a risk to the health and safety of the occupants
- cause major disruption to complete
- present a risk to the NHBC Buildmark warranty.

A new property on which a monitoring programme was ongoing would clearly not fall into the category of being completed. A verification report once prepared, would include, *inter alia*, all of the monitoring data obtained and the monitoring programme itself, would then cease.

As described above, the monitoring programme should have a defined date of cessation. The longevity of the programme should reflect the conceptual site model and so should take into account the nature of the soil gas regime, any likely natural variations (for example seasonal) etc. It should also take into account the practical aspects outlined above such as:

1 Legal/contractual requirements for completion/sale.
2 The opportunity for monitoring during construction of the above ground structure.
3 The phased nature of any development.
4 The potential impact on risk perception for all stakeholders.

Taking into account all of this variability, it is not practicable to define a fixed period for such a monitoring programme. However, it is suggested that for a site with a well-defined gas regime and falling into "amber" category a monitoring period of three to six months following completion of groundworks/foundation construction (of a phase of a large development, or the whole of a small development) is likely to be appropriate.

For a site where the soil gas regime is less well-defined and/or falling into a "Red" category (Note: this would not normally apply to residential development sites) this period may need to extend to one year (or longer if any uncertainty remains). Such a verification monitoring programme should be carried out in accordance with the good practice procedures described in Chapters 4 to 6.

Any such programme should be responsive to the data obtained. The objective of carrying out such a monitoring programme is not merely to obtain data. There should be a management response mechanism defined, and in place, capable of responding to the data obtained.

10.7 VERIFICATION/COMPLETION REPORTING

Verification is an essential element of any remedial strategy. The verification process aims to provide assurance that the original risk management objectives have been satisfactorily achieved and so provide confidence to all stakeholders. This requires that the process itself is transparent, robust and provides a data set demonstrating adequate performance of the remediation measures ("lines of evidence") (Defra and Environment Agency, 2004a).

If the remedial strategy includes a long-term/post construction monitoring programme (as described above) then the results of such a programme will also need to be incorporated in a final completion report, or may form complementary volumes to the initial verification report prepared at the conclusion of the remedial works.

Project completion and the demonstration of such completion will always be site-specific but will relate to the risk management strategy particular to each site (Environment Agency and NHBC, 2000). The typical contents of a verification report are described in the Model Procedures (Defra and Environment Agency, 2004a).

11 Recommendations for research

1. There is a lack of research data and practical experience regarding the relevance, accuracy and interpretation of results obtained from specialist monitoring techniques (eg venting and re-circulation). It is recommended that further research into these specialist monitoring techniques is undertaken.

2. The use of flux boxes to determine surface emission rates is not well understood. This guide has highlighted the potential for significant errors in the data set, in particular when defining the chamber as static. Further research into the use of such chambers and the situation the chamber represents is recommended.

3. Notwithstanding the techniques already available for the measurement of flow rate (as described in this guide) there remains considerable uncertainty regarding their relative ease of use, consistency, reliability and interpretation. Further research into the measurement and interpretation of borehole flow rates is recommended.

4. Further research into the relationship between borehole flow rate and the surface emission rate is also recommended.

5. It is currently unknown, whether small diameter installations (eg window sample holes) provide gas data closer to actual soil concentrations in the ground than data recorded from larger diameter boreholes (eg 150 mm). Further research is recommended into the use of smaller diameter installations for the measurement of soil gases and the comparison of the results against those collected from larger diameter boreholes.

6. Current soil gas monitoring techniques typically comprise readings at intervals (eg weekly) from boreholes. However, the extent to which this current monitoring technique provides data which can accurately record the variability of gas concentrations in the ground over time is not known. Further research into continuous in borehole gas monitoring equipment is required. Work will also be required to understand the relationship between data generated by such facilities and data obtained at intervals.

7. Further research into the use of soil incubation laboratory testing for the measurement of gas production potential of a soil is required. For further information, see Kitcherside and Webster (2000) and Harries *et al* (1995)

8. A relatively thin layer of made ground commonly occupies the near surface of typical "brownfield" sites. This made ground has the potential to generate soil gases, which generally comprise low methane concentrations and higher carbon dioxide concentrations. However, it is unknown whether soil gas on such low risk sites actually migrates from the ground. Research into this area will enable typical low risk sites to be further classified (also see Section 2.1).

9. Further research into methane oxidation at shallow depths is required. This research will determine whether gas protection is necessary on sites where the methane is being oxidised.

10. This guide has summarised the potential for meteorological influences on the soil gas regime at a site. However, there can sometimes be very little interaction between climatic events and the gas regime. Further research into the interpretation of such interactions is required. In addition, the relationship between atmospheric pressure and soil gas pressure requires further research.

11 This guide has referred to a method for defining a "characteristic situation" for the soil gas regime at a site which incorporates "the Pecksen methodology" (1986). A number of previous authors and practitioners have raised concerns regarding the assumptions behind the Pecksen methodology, in particular the assumption of a 10 m² zone of influence of a standpipe. Further research is required into the validity of this assumption regarding this zone of influence.

12 Further research should be carried out to quantify the ventilation rates from sub-floor voids.

13 Further work into the engineering of design gas protection should be undertaken. In particular a comparison between Eurocode and British Standards is recommended.

14 This guide has focused on "typical" main components of landfill gas soil gases. Further research into the toxicity of trace components of landfill gas and the assessment of such components is recommended.

15 Further research into the durability of plastics in a soil gas environment is recommended. For example, it is unknown whether plastics deteriorate in the presence of volatile organic compounds.

16 The industry should aim to give some quantitative measures to the typical terminology used to describe the generation potential of source (Table 5.5a and 5.5b).

12 References

AERC (2001)
Landfill gas trace constituents database
Compiled by Applied Environmental Research Centre Ltd (AERC), Pre-Release Draft, April 2001

Arup (1999)
Confidential report

Association of Geotechnical and Geo-environmental Specialists (1998)
The AGS Code of Conduct for site investigation
AGS <http://www.ags.org.uk/businesss/conductcode.cfm>

Association of Geotechnical and Geoenvironmental Specialists (2000)
Guidelines for combined geoenvironmental and geotechnical investigations
AGS <http://www.ags.org.uk/publications/pubcat.cfm#combined%20SI>

ASTM (2002)
Standard/guide for risk-based corrective action applied at petroleum release sites
ASTM Standard E2081-00(2004)e1

BANNON, M P and HOOKER, P J (1993)
Methane: Its occurrence and hazards in construction
Report 130, CIRIA, London

BARRY, D L (1986)
"Hazards from methane on contaminated sites"
In: *Proc Int Conf on Building on marginal and derelict land, Glasgow, 7–9 May, pp 209–224.*
Report 130, CIRIA, London

BARRY, DL, SUMMERSGILL, IM, GREORY, RG and HELLAWELL, E (2001)
Remedial engineering for closed landfill sites
C557, CIRIA, London

BOLTZE, U and DE FREITAS M H (1996)
"Changes in atmospheric pressure associated with dangerous emissions from gas generating disposal sites. The "explosion risk threshold" concept"
Geotechnical Engineering, **119**, pp 177–181

BOYLE, R and WITHERINGTON, P (2007)
Guidance on evaluation on development proposals on sites where methane and carbon dioxide are present, incorporating "traffic lights"
Report Ref 10627-R01-(02), National House Building Council

BRE (1999)
Radon: Guidance on protective measures for new dwellings
BRE Report 211, BRE Press, Berkshire (ISBN: 1-86081-328-3)

BRE (1995c)
Positive Pressurisation: a BRE guide to radon remedial measures in existing dwellings
BRE Report 281, BRE Press, Berkshire (ISBN: 1-86081-007-1)

BRE (1999)
Radon: Guidance on protective measures for new dwellings
3rd edn, BRE Report 211, BRE Press, Berkshire (ISBN: 1-86081-328-3)

CARD, G B (1996)
Protecting development from methane
Report 149, CIRIA, London

CREEDY, D, SCEAL, J and SIZER, K (1996)
Methane and other gases from disused coal mines: the planning response technical report
Wardell Armstrong, DoE, London, Stationery Office

CROWHURST, D (1987)
Measurement of gas emissions from contaminated land
BRE Press, Watford (ISBN: 0-85125-246-X). Now out of print

CROWHURST, D and MANCHESTER, S J (1993)
The measurement of methane and other gases from the ground
CIRIA Report 131, CIRIA, London

DAWSON, H E (2002)
"Evaluating vapour intrusion from groundwater and soil to indoor air"
In: Proc EPA Brownfields Conference, Charlotte, 13 November. Available from:
<www.epa.gov/correctiveaction/eis/vapor/f02052.pdf>

Defra (2004)
Review of health and environmental effects of waste management. Phase 1 – Municipal solid waste and similar wastes
Product code PB9052A, Department for Environment Food and Rural Affairs. Available from: <www.defra.gov.uk/environment/waste/research/health/pdf/health-report.pdf>

Defra and EA (2002a)
Assessment of risks to human health from contamination: An overview of the development of soil guideline values and related research
R&D Publication CLR7, Department for Environment Food and Rural Affairs and Environment Agency

Defra and EA (2002b)
Potential contaminants for the assessment of land
R&D Publication CLR8, Department for Environment Food and Rural Affairs and the Environment Agency

Defra and EA (2002c)
Contaminants in soil: Collation of toxicological data and intake values for humans
R&D Publication CLR9, Department for Environment Food and Rural Affairs and the Environment Agency

Defra and EA (2002d)
The Contaminated Land Exposure Assessment Model (CLEA): Technical basis and algorithms
R&D Publication CLR10, Department for Environment Food and Rural Affairs and the Environment Agency

Defra and the EA (2004a)
Model Procedures for the management of land contamination
Contaminated Land Report 11, Department for Environment Food and Rural Affairs and the Environment Agency, Bristol

Defra and the EA (2004b)
Soil guideline values for toluene contamination
Science Report SGV 15, Department for Environment Food and Rural Affairs and Environment Agency, Bristol

Defra and EA (2005a)
Soil guideline values for Ethyl benzene contamination
Science Report SGV 16, Department for Environment Food and Rural Affairs and the Environment Agency, Bristol

DoE (1991)
Landfill gas
Waste Management Paper No 27, The Stationary Office, London

DoE (1994a)
Guidance on preliminary site inspection of contaminated land
Contaminated Land Research Report No 2, Department of Environment

DoE (1994b)
Documentary research on industrial sites
Contaminated Land Research Report No 3, Department of Environment

DoE (1994c)
Sampling strategies for contaminated land
Contaminated Land Research Report No 4, Department of Environment, 1994

Department of the Environment, Transport and the Regions (2000b)
Guidelines for environmental risk assessment and management
The Stationery Office, Norwich

Department of the Environment, Transport and the Regions and Partners in Technology (1997)
Passive venting of soil gases beneath buildings, guide for design
Research Report, Volume I, Ove Arup and Partners

DERWENT, R G; JENKIN, M E and SANDERS, S M (1996)
Photochemical ozone creation potential for a large number of reactive hydrocarbons under European conditions
Atmospheric Environment 30, **2**, pp181–199

EARL, N (2003)
"Technical aspects of SGVs and application of the CLEA model at contaminated land: Soil guideline values, progress and prognosis"
In: *Proc Society of Chemical Industry, London, 29 April*

EARL, N; CARTWRIGHT, C D; HORROCKS, S J; WORBOYS, M; SWIFT, S; KIRTON, J A; ASKAN, A U; KELLEHER, H and NANCARROW, D J (2003)
Review of the fate and transport of selected contaminants in the soil environment
Environment Agency Draft Technical Report P5-079/TR1, September 2003

Environment Agency (2002c)
GasSim –landfill gas risk assessment tool
R&D Project P1-295, Environment Agency, Bristol <www.gassim.co.uk>

EA (2003a)
Guidance on landfill completion: A consultation
Environment Agency, Bristol

EA (2003b)
Consultation on agency policy: Building development on or within 250 metres of a landfill site. Background information
A consultation, Environment Agency, Bristol

EA (2003c)
HI Environmental assessment and appraisal of best technology (Version 6)
Environment Agency, Bristol

EA (2004a)
Guidance on the management of landfill gas
LFTGN 03, Environment Agency, Bristol

EA (2004b)
Guidance for monitoring trace components in landfill gas
LFTGN 04, Environment Agency, Bristol

EA (2004c)
Update on estimating vapour intrusion into buildings
CLEA Briefing Note 2. Available from: <www.environment-agency.gov.uk/commondata/105385/soil_vapour_intrusion_749183.pdf> (accessed 10 June 2004)

EA (2004d)
The likely medium to long-term generation of defects in geomembrane liners
R&D Technical Report P1-500/1TR, Environment Agency, Bristol

EA (2005a)
Review of building parameters for the development of a soil vapour intrusion model
Environment Agency, Bristol

EA (2005b)
Update of supporting values and assumptions describing UK building stock
CLEA Briefing Note 3, Environment Agency, Bristol. Available from: <www.environment-agency.gov.uk/commondata/acrobat/bn3_904797.pdf> (accessed 19 January 2005)

EA and NHBC (2000)
Guidance for the safe development of housing on land affected by contamination
R&D Publication 66, Environment Agency, Bristol

EA, Environment and Heritage Service, and SEPA (2002a)
Horizontal guidance for odour Part 1: Regulation and permitting
Technical Guidance Note IPPC H4, Draft. Environment Agency, Bristol

EA, Environment and Heritage Service, and SEPA (2002b)
Horizontal guidance for odour Part 2: Assessment and control
Technical Guidance Note IPPC H4, Draft. Environment Agency, Bristol

EHRIG, H (1996)
Prediction of gas production for laboratory-scale tests landfilling of waste: Biogas
Christensen Cossu (ed), Stegmann Spon

EVANS, D; HERS, I, WOLTERS, R M, BODDINGTON, R T B and HALL, D H (2002)
Vapour transport of soil contaminants
P5-018/TR, Research and Development Technical Report, Environment Agency and Institute of Petroleum

FERGUSON, C and KRYLOV, V (1998)
"Contamination of indoor air by toxic soil vapours: the effects of subfloor ventilation and other protective measures"
Building and Environment, Vol 33, **6**, November, Elsevier, pp 331–347

FERGUSON, C; NATHANAIL, P; McCAFFREY, C; EARL, N; GILLET, A and OGDEN, R (2003)
Method for deriving site specific human health assessment criteria for contaminants in soil report No LQ01
SNIFFER, Edinburgh <www.sniffer.org.uk>

FISHER, L J, ABRIOLA, L M, KUMMLER, R H, LONG, D T and HARRISON, K G (2001)
Evaluation of the Michigan Department of Environmental Quality's generic groundwater and soil volitization to indoor air inhalation criteria
Michigan Environmental Science Board, Lansing

Glasgow Scientific Laboratories (2005)
Correspondence

GODSON, J A E and WITHERINGTON, P J (1996)
Evaluation of risk associated with hazardous soil gases
Fugro Environmental, Manchester

GREGORY, R G, REVANS, A J, HILL, M D, MEADOWS, M P, PAUL, L and FERGUSON, C C (1999)
A framework to assess the risks to human health and the environment from landfill gas
Technical Report P271, under contract CWM 168/98, Environment Agency, Bristol

HARRIES, C R, McENTEE, J M and WITHERINGTON, P J (1995)
Interpreting measurements of gas in the ground
Report 151, CIRIA, London

HARRIS, M R, HERBERT, S M and SMITH, M A (1996)
Remedial treatment for contaminated land, Volume VI: Containment and hydraulic measures
Special Publication 106, CIRIA, London

HARTLESS, R (1991)
Construction of new buildings on gas-contaminated land
BRE Report 212, BRE Press, Berkshire (ISBN: 0-85125-513-2)

HARTLESS, R P (2000)
"Developing a risk assessment framework for landfill gas – incorporating meteorological effects"
In: *Proc Conf Waste 2000: Waste Management at the Dawn of the Third Millennium, 2–4 October, Stratford, pp41–50*

Health Protection Agency
<www.hpa.org.uk>

Health Protection Agency
<www.hpa.org.uk/radiation>

HERS, I (2000)
"Measurement of BTX vapour intrusion into an experimental building"
In: *Presentation at the United States Environmental Protection Agency (USEPA) Resource Conservation and Recovery Act (RCRA) Corrective Action Environmental Indicator Forum, 15 August 2000.* Available from: <clu-in.com/eiforum2000/prez/22/22.pdf>

HOOKER, P J and BANNON, M P (1993)
Methane: Its occurrence and hazards in construction
Report 130, CIRIA, London

HRUDEY, S E (1996)
A critical review of current issues in risk assessment and risk management
Environmental Risk Management Paper ERC 96-8, University of Alberta, Canada

HSE (2002)
Occupational exposure limits
EH40/92 1992, HMSO, London

ICRCL (1987)
Guidance on the assessment and redevelopment of contaminated land
2nd edn, Guidance Note 59/83, Interdepartmental Committee on the Redevelopment of Contaminated Land, July, DoE

ICE (1992)
Nomenclature for hazard and risk assessment
Institution of Chemical Engineers

Institute of Petroleum (1998)
Guidelines for investigation and remediation of petroleum retail sites
The Institute of Petroleum, London

Institutes of Waste Management (1998)
The monitoring of landfill gas
2nd edn, Institutes of Waste Management

Intergovernmental Panel on Climate Change (1996a)
Guidelines for national greenhouse gas inventories reference manual, Chapter 6
Revised 1996 IPCC

Intergovernmental Panel on Climate Change (1996b)
IPCC Expert group on waste, topical workshop on carbon conversion and methane oxidation in solid waste disposal sites
Argonne National Laboratory, Chicago, USA, 25 October 1996

ITRC and Brownfields Team (2003)
Vapour intrusion issues at brownfield sites
Interstate Technology and Regulatory Council, and Brownfields Team

ITRC and Brownfields Team (2004)
Regulatory acceptance for new solutions: vapour intrusion (indoor air)
Interstate Technology and Regulatory Council, and Brownfields Team. Available from: <www.itrcweb.org>

JOHNSON, R (2001)
Protective measures for housing on gas-contaminated land
BRE Report 414, BRE Press, Berkshire (ISBN: 1-86081-460-3)

JOHNSON, P C and ETTINGER, R A (1991)
"Heuristic model for predicting the intrusion rate of contaminant vapours into buildings"
Environmental Science and Technology, **25,** pp 1445–1452

JOHNSON, P C, ETTINGER, R A, KURTZ, J, BRYAN, R and KESTER, J E (2002)
Migration of soil gas vapours to indoor air: determining vapour attenuation factors using a screening-level model and field data from the CDOT-MTL, Denver, Colorado site
Bulletin 16, American Petroleum Institute

KANOL, D W and ZETTHER, G H (1990)
In: *Proc of the 5th Sewage and Refuse Symposium*
Abwassertechnische Vereinnigung ev, Munich, 198, Report 130, CIRIA, London

KENNEDY, J; MARNICIO, R; CARAVATI, M and KENNEDY, E (1998)
RBCA Fate and transport models: Compendium and selection guidance
Report by Foster Wheeler Environmental Corporation to the American Society for Testing and Materials, ASTM, November 1998

KING, P J; MUNDAY, G and RYAN G (1988)
"Report of the non-statutory public inquiry into the gas explosion at Loscoe, Derbyshire 24 March 1986"

KRYLOV, V V and FERGUSON C C (1998)
"Contamination of indoor air by toxic soil vapours: the effects of sub-floor ventilation and other protective measures"
Building and Environment, 33, **6**, pp 331–347

LIZJEN, J P A, BAARS, A J, OTTE, P F, RIKKEN, M G J, SWARTIES, F A, VERBRUGGEN, E M J, and VAN WEZEL A P (2001)
Technical evaluation of the intervention values for soil/sediment and groundwater; human and ecotoxicological risk assessment and derivation of risk limits for soil, sediment and groundwater
RIVM Report 711701023, The Netherlands National Institute of Public Health and the Environment

ODPM (2004a)
Approved Document C: Site preparation and resistance to contaminants and moisture
The Stationary Office, Norwich (ISBN: 978-1-85946-202-7)
<www.planningportal.gov.uk/uploads/br/BR_PDFs_ADC_2004.pdf>

ODPM (2004b)
Planning Policy Statement 23: *Planning and pollution control*
The Stationary Office, Norwich

O'RIORDAN, N J and MILLOY, C J (1995)
Risk assessment for methane and other gases from the ground
Report 152, CIRIA, London

OTTE, P F, LIZJEN, J P A, OTTE, J G, SWARTJES, F A and VERSLUIJS, C W (2001)
Evaluation and revision of the CSOIL parameter set; proposed parameter set for human exposure modelling and deriving intervention values for the first series of compounds
RIVM Report 711701021, The Netherlands National Institute of Public Health and the Environment

OWEN, R and PAUL, V (1998)
Gas protection measures for buildings, methodology for the quantitative design of gas dispersal layers
Ove Arup & Partners

PECKSEN, G N (1986)
Methane and the development of derelict land
London Environmental Supplement, Summer 1985, No 13, London Scientific Services, Land Pollution Group

POLETTI, E, HAYWARD, H, GILL, J, BAKER, K, GARDNER, M, HOULDEN, L (in press)
Risk assessment comparison study
Arcadis Geraghty & Miller International Incorporated Final Draft Report 916830024 to NICOLE/ISG (Network for Industrially Contaminated Land in Europe Industrial Sub-Group), July 2003

POLSON, C J and MARSHALL, T K (1975)
The disposal of the dead
3rd edn, The English Universities Press Ltd

PRIVETT, K D, MATTHEWS, S and HODGES, R A (1996)
Barriers, liners and cover systems for containment and control of land contamination
Special Publication 124, CIRIA, London

RAYBOULD, J G, ROWAN, S P and BARRY, D L (1995)
Methane investigation strategies
Report 150, CIRIA, London

RIKKEN, M G J, LIZJEN, J P A and CORNELESS, A A (2001)
Evaluation of model concepts on human exposure
RIVM Report 711701022, The Netherlands National Institute of Public Health and the Environment

RILEY, M R, JORDAN, K A and COX, M L (2004)
"Development of a cell-based sensing device to evaluate toxicity of inhaled materials"
Biochemical Engineering Journal, 19, pp 95–99

RUDLAND, D J, LANCEFIELD, R M and MAYELL, P N (2001)
Contaminated land risk assessment
C552, CIRIA, London

SCHUVER, H J (2003)
"Overview of the science behind USEPA's guidance for the vapour intrusion to indoor air pathway"
In: *USEPA Indoor Air Vapour Intrusion Seminar, 14 January 2003*

SCIVYER, C R and GREGORY, T J (1995)
Radon in the workplace
BRE Report 293, BRE Press, Berkshire (IBSN: 1-86081-040-3)

Site Investigation Steering Group (1993)
Guidelines for the safe investigation by drilling of landfills and contaminated land
Thomas Telford, London

SLADEN, J A, PARKER, A and DORELL, G L (2001)
"Quantifying risks due to soil gas on brownfield sites"
Land Contamination and Reclamation, 9, **2**, EPP Publications, London

SPENCE, L R and WALDEN, T (2001)
"Risk integrated software for clean-ups"
User's manual, Version 4.0, BP Global Environmental Management

STEEDS, J E, SHEPHERD, E and BARRY, D L (1996)
A guide for safe working on contaminated sites
Report 132, CIRIA, London

STEVENS, R and CROWCORFT, P (1995)
"Positive pressure air systems for protection of buildings from soil gas ingress – a case study"
Land Contamination and Reclamation, 3, **1**, EPP Publications, London

STODDART, J; ZHU, M; STAINES, J; ROTHERY, E and LEMICKI, R (1999)
"Experience with halogenated hydrocarbons removal from landfill gas"
In: *Proc Sardinia 1999, Seventh International Waste Management Landfill Symposium, S Margherita di Pula, Cagliari, Sardinia, 4–8 October, Volume II, pp 489–498*. CISA, Cagliari, Italy

SWARTJES, F A (2001)
"Human exposure model comparison study: state of play"
Land Contamination and Reclamation, 9, **1**, pp 101–106

TEDD, P, WITHERINGTON, P, EARLE, D, HOLLINGSWORTH, S, FURLONG, B, BRADLEY, L, MALLETT, H and LAIDLER D (2004)
Cover systems for land regeneration – thickness of cover systems for contaminated land
BRE Report 465, BRE Press, Berkshire (ISBN: 1-86081-684-3)

TINDLE, P E (2002)
Method for monitoring exposure to gasoline vapour in air – revision 2002
Report No 8/02, Concawe, Brussels

USEPA (1991)
User's guide for evaluating subsurface vapour intrusion into buildings
USEPA, Washington

USEPA (1997)
Compilation of air pollutant emission factors
5th edn, AP-42, Vol 1, Chapter 2, Solid Waste Disposal, 2.4 MSW Landfills

USEPA (1997/2003)
User's Guide for the Johnson and Ettinger (1991) Model for subsurface vapour intrusion into Buildings prepared by Environmental Quality Management, Inc., Contract No. 68-D30035.
Available from: www.epa.gov/superfund/programs/risk/airmodel/johnson_ettinger.htm>

USEPA (2001c)
Evaluating the vapour intrusion to indoor air pathway
Draft supplemental guidance

USEPA (2002)
OSWER *Guidance for evaluating the vapour intrusion to indoor air pathway from groundwater and soils* (subsurface vapour intrusion guidance) Available from:
<www.epa.gov/epaoswer/hazwaste/ca/eis/vapour.htm>

WELSH, P A (1995a)
Testing the performance of terminals for ventilation systems, chimneys and flues
BRE IP5/95, BRE Press, Berkshire

WELSH, P A (1995B)
Flow resistance and wind performance of some common ventilation terminals
BRE IP6/95, BRE Press, Berkshire

WHITTAKER, J J, BUSS, S R, HERBERT, A W and FERMOR, M (2001)
Benchmarking and guidance on the comparison of selected groundwater risk assessment models
Environment Agency National Groundwater and Contaminated Land Centre Report NC/00/14

WILSON, S A and CARD, G B (1999)
"Reliability and risk in gas protection design"
Ground Engineering, 32, **2**, February, EMAP, London, pp 32–36

WILSON, S A and HAINES, S (2005)
"Site Investigation and monitoring for soil gas assessment – back to basics"
Land Contamination and Reclamation, 13, **3**, EPP Publications Ltd

ZAMPOLLI, S, ELMI, I, and AHMED, F, PASSINI, M, CARDINALI, G C, NICOLETTI, S and DORI, L (2004)
"An electronic nose based on solid state sensor arrays for low-cost indoor air quality monitoring applications"
Sensors and Actuators (B: Chemical), 101, 1–2, pp 39–46

British Standards

BS 5930:1999: *Code of practice for site investigations*

BS 10175:2001: *Investigation of potentially contaminated sites – code of practice*

Acts

Environmental Protection Act 1990: Part IIA Contaminated Land
Circular 02/2000a, Department of the Environment, Transport and the Regions, London, 2000

A1 Trace components

Significant trace component	Sampling method
Chloroethane	Dual solid sorbent
Chloroethene (vinyl chloride)	Dual solid sorbent
Benzene	Dual solid sorbent
2-butoxy ethanol	Dual solid sorbent
Arsenic (as As)	Solid sorbent
1,1-dichloroethane	Dual solid sorbent
Trichloroethene	Dual solid sorbent
Tretrachloromethane	Dual solid sorbent
Methane (formaldehyde)	Reactive sorbent
Hydrogen sulphide	Direct on site measurement of raw gas
1,1-dichlorethene	Dual solid sorbent
1,2-dichloroethene	Dual solid sorbent
Carbon disulphide	Dual solid sorbent
Methanethiol	Dual solid sorbent
Butyric acid	Dual solid sorbent
Ethanal (acetaldehyde)	Reactive sorbent
Ethyl butyrate	Dual solid sorbent
1-propanethiol	Dual solid sorbent
Dimethyl disulphide	Dual solid sorbent
Ethanethiol	Dual solid sorbent
1-pentene	Dual solid sorbent
1-butanethiol	Dual solid sorbent
Dimethyl sulphide	Dual solid sorbent
1,3-butadiene	Dual solid sorbent
Furan	Dual solid sorbent
Mercury (as Hg)	Solid sorbent

A2 Aquifer protection

Aquifer protection measures, to reduce the potential for any downward migration of contaminants, should be utilised during borehole construction where there is a potential to create to a pathway. Typically boreholes begin at a larger diameter (eg 50 mm larger than the final drilled diameter for each protection measure that is required). A bentonite plug is then installed at the base of the potentially contaminated strata, ideally, coincident with low permeability strata. Drilling continues at a reduced diameter (usually by 50 mm) to the elevation of the next seal or the base of the borehole.

The bentonite plug should be formed by placing bentonite pellets or powder in the base of the borehole to form a seal at least 1 m thick. Sufficient time should be allowed for the bentonite to swell. This can be helped by hydrating the bentonite before installation but time should still be allowed for the seal to form.

On completion a length of slotted casing, with a geotextile wrap if appropriate, should be installed at the base of the boreholes. The maximum extent of the slots should be to just below (0.2 m+) the base of the bentonite plug. The borehole should be backfilled with a gravel filter surrounding the slotted section of pipe, a sand block and bentonite plug on top of the gravel filter. The borehole should then be completed with cement bentonite grout and concrete to ground level.

A3   Soil gas monitoring proforma

Site reference no: _____

Client: _____

Site: _____

National Grid reference: _____

Date/time of monitoring: _____

Monitoring personnel: _____

Instrument type: _____

Serial number: _____

Barometric pressure
(including any trend in rise or
fall of pressure) (uncorrected): _____

Air temperature (°C): _____

Recent weather (eg precipitation,
wind speed) _____

Ground conditions
(including vegetation stress,
visual contamination) _____

Record any other observations which may have an impact on the soil gas monitoring results. These observations may include damage to the gas tap or top of well, damage to the cover or an open gas tap.

....

....

....

Geotechnical Instruments Analox GA94/2000

Typical accuracy (manufacturer supplied):

Concentration (% v/v)	CH4	CO2	O2
5 % (LEL CH4)	+/- 0.5 %	+/- 0.5 %	+/- 1.0 %
15 % (UEL CH4)	+/- 1.0 %	+/- 1.0 %	+/- 1.0 %
100 %	+/- 3.0 %	+/- 3.0 %	+/- 1.0 %
Permitted temperature range		-10°C to +40°C	

Note: Calibration records held in equipment register

Site:

Project No:

Date:

Monitoring point reference	Flow range (litres/hr)	Atmospheric pressure range (Pascals)	Methane % v/v		Methane % LEL		Carbon dioxide % v/v		Oxygen % v/v		Water level (mbgl)	Depth of well (m)	Vol of gas in well (m³)	Other gases (eg hydrogen sulphide, carbon monoxide)	Notes
			Peak	Steady	Peak	Steady	Peak	Steady	Peak	Steady					

LEL - Lower Explosive Limit

Initial

A4　Available laboratory-based gas analysis

Analysis	Principle	Gases	When to use
Gas concentration measurements			
Gas chromatography (CG-MS)	Sample is injected into a gas stream, which results in each compound separating. Each compound leaves the stream at a specific time. As it leaves the stream, it passes through a detector. The response of the detector is related to the concentration of the compound. By comparing the response of the detector to known standards, the concentration of the soil gas in the sample can be calculated. The mass spectrometer is a type of detector for a gas chromatograph. The detector separates the compounds by subjecting them to an ionising field. Each compound has a specific pattern of ions and a computer is used to identify these compounds.	Bulk gas/ trace components	This method is recommended and it is typically the most common technique. The results are accurate and the gas can be determined quantitatively and qualitatively. The analysis can be completed within 24 to 48 hours of receipt provided the work is pre-scheduled. However, the value of the results depends on the competence of the sampling method.
Portable gas chromatography (GC-MS)	With advances in technology the GC-MS has been reduced in size such that it can now be used as field equipment. Weighing between 20 kg and 35 kg the CG-MS is robust and portable.	Bulk gas/trace components	This relatively new equipment offers the accuracy of the laboratory version but also mobility, so when transported to site, the GC-MS can be ready to analyse samples within 30 minutes. However, to set up such equipment is expensive and would only be cost effective for long-term monitoring.
ICP-MS	Inductively coupled plasma mass spectrometry is a multi-element technique for the determination of ultra-trace levels of analytes in a variety of sample matrices. The argon ICP generates singly charged positive ions from the aspirated sample. These ions are separated on their mass to charge ratio by a quadruple mass spectrometer and detected (EA, 2004).	Trace components	This method is relatively new and very sensitive. Ideal for the detection of trace components at low concentrations (parts per trillion).
Gas age and formation measurements			
Carbon 14 dating	Carbon 14 is a naturally occurring radioactive isotope with a half life of about 5700 years. Carbon 14 dating is applicable to any organic matter. All sources of methane are derived from plant matter either directly (for example marsh gas) or indirectly (for example LFG). In turn all carbon present in plants is obtained by photosynthesis from atmospheric carbon dioxide which contains a small proportion of the carbon 14 isotope. The carbon 14 isotope is formed naturally in the upper layers of the atmosphere, so that once fixed by photosynthesis in the plant it decays radioactively with no mechanism for replacement. The carbon 14 content will reduce from the value equivalent to that in atmospheric carbon dioxide in a conventional radioactive decay curve with a half life of 5700 years. So, it follows that mine gas and mains gas being geologically ancient have a very small carbon 14 content whereas that present in LFG and other recent sources is substantial.	Methane and carbon dioxide	This method requires large volumes of gas and sophisticated analytical facilities which only a few laboratories can provide in the UK.
Stable isotope measurements	Samples are placed in clean metal capsules and loaded into an automatic sampler. They are then dropped into a furnace held at 1000°C where they are combusted in the presence of added oxygen. The metal capsules are flash combusted, raising their temperature in the region of the sample to ~ 1700°C. The combusted gases are then swept in a helium stream over a combustion catalyst (Cr2O3), CuO wires (to oxidise hydrocarbons), and silver wool to remove sulphur and halides. The remaining gases, N2, NOx, H2O, O2, and CO2 are then swept through a reduction stage of pure copper wires held at 600°C. This step removes any oxygen and converts NOx to N2. Water is removed by a magnesium perchlorate while CO2 can be removed via a selectable Carbosorb trap. Nitrogen and carbon dioxide are separated by packed column gas chromatograph held at an isothermal temperature. The resultant chromatographic peak enters the ion source where it is ionised and accelerated. Gas species of different mass are separated in a magnetic field then simultaneously measured <www.iso-analytical.com/page17.html>.	Methane, hydrocarbon vapours	This method is fairly new. Only a few laboratories can provide in the UK.

A5 Quantitative risk assessment

A5.1 INTRODUCTION

To provide a quantitative analysis of risk the following questions should be answered:

1 **What can go wrong?**
Identify combination of events that can create an undesirable event. For soil gas, examples of the undesirable event include an explosion in a building, asphyxiation in a confined space or methane causing die back of vegetation in an adjacent nature reserve. This list is not exhaustive and there will be other examples. The events that lead to an event of concern can be wide ranging and many of those related to landfill gas are difficult to quantify. Examples of events can include accidental damage of a gas membrane by an occupier or blocking of underfloor vents by constructing a conservatory. Again these are examples and a site-specific risk register should be developed for every assessment.

2 **How likely is it?**
The terms "likelihood", "probability" and "frequency" are all used in relation to risk assessment and it is important that they are clearly understood and used in a consistent way.

   – *Likelihood* is an expression which indicates, in general terms, the possibility of something happening.

   – *Probability* is the number of favourable outcomes divided by the total number of possible outcomes of an experiment. It is expressed as a number in the range zero to one; zero being the certainty that an event will not occur, and one the certainty that an event will occur. Probability can also be expressed in percentage terms.

   – *Frequency* expresses how often an event occurs within a given time. It is defined as the reciprocal of the average time between events, and is often expressed in terms such as one per 10 000 years. The difference between frequency and probability is clearly demonstrated by noting that for an event which is likely to occur every three months, the frequency of occurrence is four per year whereas the probability of the event occurring in any one year will be somewhere close to a value of one.

3 **What are the consequences?**
Consequence is a measure of the magnitude of the effects of an event. This can be expressed as the number of fatalities or as a monetary value. The nature and severity of the consequences of an event determine the acceptable risk of the event occurring. Greater severity consequences are less acceptable, see Table A5.1.

The overall risk is defined by the Institution of Chemical Engineers (1992) as:

Risk = Frequency × Consequences

(where frequency is in units of time and consequences are measured in terms of number of injuries, number of fatalities or a monetary value of loss).

The method of quantitative risk assessment proposed in CIRIA publication R152 *Risk assessment for methane and other gases from the ground* (O'Riordan *et al*, 1995) is a very simplified approach to quantitative risk assessment that has been adopted in the UK for assessing risks relating to soil gas. It does not consider the effects of an explosion and in other industries such as offshore oil and gas production there are complex methods of assessing the consequences of an explosion. At present these are not used for soil gas risk assessments as any explosion is considered undesirable.

Table A5.1 *Acceptability of risk*

Event causing total loss					
		Annual likelihood		Attitude to reliability	
Degree of risk	To life	To property	To money	Voluntary risk	Involuntary risk
Very risky	1 in 100	1 in 10	1 in 1	Very concerned	Totally unacceptable
Risky	1 in 1000	1 in 100	1 in 10	Concerned	Not acceptable
Some risk	1 in 10 000	1 in 1000	1 in 100	Circumspect	Very concerned
A slight chance	1 in 100 000	1 in 10 000	1 in 1000	Of little concern	Concerned
Unlikely	1 in 1 million	1 in 100 000	1 in 10 000	Of no concern	Circumspect
Very unlikely	1 in 10 million	1 in 1 million	1 in 100 000	Of no concern	Of little concern
Practically impossible	1 in 100 million	1 in 10 million	1 in 1 million	Of no concern	Of no concern
Event causing impairment					
		Annual likelihood		Attitude to reliability	
Degree of risk	To life	To property	To money	Voluntary risk	Involuntary risk
Very risky	1 in 10	1 in 1	10 in 1	Very concerned	Not unacceptable
Risky	1 in 100	1 in 10	1 in 1	Concerned	Very concerned
Some risk	1 in 1000	1 in 100	1 in 10	Circumspect	Concerned
A slight chance	1 in 10 000	1 in 1000	1 in 100	Of little concern	Circumspect
Unlikely	1 in 100 000	1 in 10 000	1 in 1000	Of no concern	Of little concern
Very unlikely	1 in 1 million	1 in 100 000	1 in 10 000	Of no concern	Of no concern
Practically impossible	1 in 10 million	1 in 1 million	1 in 100 000	Of no concern	Of no concern
Event causing inconvenience					
		Annual likelihood		Attitude to reliability	
Degree of risk	To life	To property	To money	Voluntary risk	Involuntary risk
Very risky	1 in 1	10 in 1	100 in 1	Very concerned	
Risky	1 in 10	1 in 1	10 in 1	Very concerned	
Some risk	1 in 100	1 in 10	1 in 1	Circumspect	
A slight chance	1 in 1000	1 in 100	1 in 10	Of little concern	
Unlikely	1 in 10 000	1 in 1000	1 in 100	Of no concern	
Very unlikely	1 in 100 000 million	1 in 10 000	1 in 1000	Of no concern	
Practically impossible	1 in 1 million	1 in 100 000	1 in 10 000	Of no concern	

Note: The acceptability of risk in the preceding tables is defined in terms of likelihood or probability and not frequency.

The Health & Safety Executive apply the following principles to risk assessment and determining what is considered to be acceptable:

Principle 1

HSE starts with the expectation that suitable controls must be in place to address all significant hazards and that those controls, as a minimum, must implement authoritative good practice irrespective of situation-based risk estimates.

Principle 2

The zone between the unacceptable and broadly acceptable regions is the tolerable region. Risks in that region are typical of the risks from activities that people are prepared to tolerate in order to secure benefits in the expectation that:

- the nature and level of the risks are properly assessed and the results used properly to determine control measures
- the residual risks are not unduly high and kept as low as reasonably practicable (the ALARP principle)
- the risks are periodically reviewed to ensure that they still meet the ALARP criteria, for example, by ascertaining whether further or new controls need to be introduced to take into account changes over time, such as new knowledge about the risk or the availability of new techniques for reducing or eliminating risks.

Principle 3

"Both the level of individual risks and the societal concerns engendered by the activity or process must be taken into account when deciding whether a risk is acceptable, tolerable or broadly acceptable". In addition the fact that "hazards that give rise to …. individual risks also give rise to societal concerns and the latter often play a far greater role in deciding whether risk is unacceptable or not" must be considered.

A5.2 FAULT TREE ANALYSIS

Fault tree analysis is the most commonly applied method for undertaking quantitative gas risk assessments in the UK and the reasons for this are given in CIRIA publication R152 (O'Riordan *et al*, 1995). Fault tree analysis is a graphical method that provides a systematic description of potential failures and events in a system that can combine to cause an unwanted event. The failures can occur due to any reason including equipment or human failures.

Fault trees are constructed by identifying the conditions that are necessary for a top event to occur (ie the critical event). The conditions for each successive underlying chain of events are identified until the basic causes are identified and the fault tree cannot be extended any further. The events are connected by logic gates. Fault tree analysis is a complex and specialist field, and anyone undertaking an analysis should be fully conversant with the application of the technique and be aware of how it is applied in other fields. The approach used for soil gas is greatly simplified and anyone undertaking an assessment should understand the implications of this. A few of the most basic symbols are shown in Figure A5.1.

Most fault tree analyses for gas risk assessment can be carried out using the following symbols only (see Figure A5.1):

- **TOP event** – a foreseeable, undesirable event that is being assessed
- **Intermediate event** – events that must occur in order for the top event to occur
- **OR gate** – produces output if any input exists.
- **AND gate** – produces output if all inputs co-exist.
- **Basic event** – A basic event is one which can not be broken down any further into component parts. Leaf and initiation are different terms of a basic event.

Steps in fault tree analysis

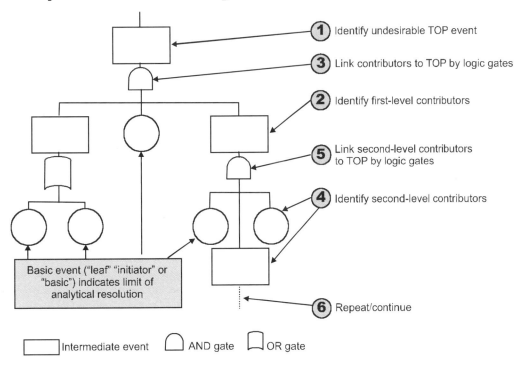

Figure A5.1 *Developing a fault tree analysis*

An example of a simple fault tree for asphyxiation in a building is provided in Figure A5.2.

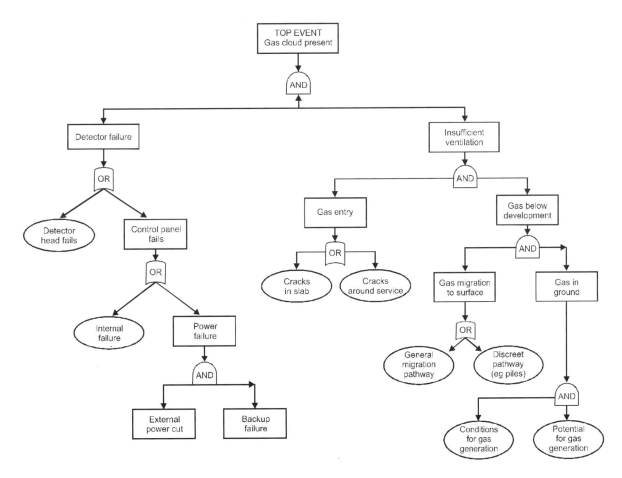

Figure A5.2 *Example of a simple fault tree for a gas cloud that could cause asphyxiation*

An important concept of fault tree analysis is that all the events and sub events are independent of each other. For example in the fault tree from CIRIA publication R152 (O'Riordan and Milloy, 1995), the cause of gas entry cannot be dependent on a factor that also causes ignition.

The first step in a fault tree analysis is to determine the possible top events for failure of the gas protective system. A common fault is to analyse only one event when several should be assessed. This should be site-specific and possible events can include:

- explosion due to a gas cloud
- presence of a gas cloud that could cause a methane explosion (both internal and external)
- presence of a gas cloud that could cause asphyxiation or acute poisoning by gas (eg carbon dioxide, hydrogen sulphide, hydrogen cyanide or carbon monoxide)
- entry of explosive or asphyxiating gas into excavations during construction.

Note that these are events and not the consequence of an event such as an explosion or death. These are dealt with outside the fault tree. These are all acute (short-term) risks. Chronic (long-term) health risks from the soil gas or vapour should be assessed using other methods, such as the contaminated land exposure assessment (CLEA) framework.

Once the fault tree has been developed probabilities or frequencies can be applied to each sub event so the overall probability or frequency of failure of the gas protective system (including any natural or other barriers such as concrete slabs) can be estimated. The difficulty in undertaking fault tree analysis for gas problems is in assigning probabilities or frequencies to some events, for example the risk of accidental damage to a membrane after it is installed. These issues are discussed in more detail in the following sections.

CIRIA publication R152 (O'Riordan and Milloy, 1995) includes ignition in the fault tree as an annual frequency (in the example it is 50 times per year) and the fault tree is used to estimate the frequency of an explosion (not the probability). At present there are differing views within the soil gas industry regarding the use of ignition frequency within the fault tree (eg Hartless, 2004, proposes that the ignition should be represented as a probability that is effectively 1 in virtually all situations). A frequency for the ignition is estimated and placed in the fault tree. The fault tree is used to calculate the frequency of an explosion rather than the annual probability. There are, however, references discussing fault tree analysis in other industries that use failure rates or frequencies within a fault tree (giving values greater than 1), and also support the approach used in CIRIA publication R152.

Further research is required on the application of fault tree assessment to soil gas problems but it is clear that the two approaches described below are estimating different parameters (one is estimating frequency, the other probability). For this reason it is important that, whichever approach is adopted, the results should not be taken as an absolute value of risk and should only be used for comparative purposes to aid judgement as described in CIRIA publication R152. It is also important that the criteria against which risk is assessed are also comparable to the calculated parameter (ie do not compare calculated frequencies with acceptable probability).

The following approach uses the frequency of ignition within the calculations, rather than probability.

Once the failure of the protective system has been assessed the frequency of an explosion or asphyxiation, FE, is based on two factors:

1. Frequency of exposure to the hazard (eg how many times an operation is performed, for example ignition).
2. The probability of experiencing a gas protective system failure during any exposure (obtained from the fault tree).

An estimate of the frequency of the event can be obtained from:

Frequency of adverse event = frequency of exposure × probability of system failure during exposure

To calculate the probability of an explosion using the approach described by Hartless (2004) the probability of ignition should be calculated as follows.

A light being switched on in a cupboard is assumed to be the ignition source in this example (but is not the only possible source). This is likely to be a random event. If this event occurs once a week on average then there is a probability that in a given week it will not be turned on at all, or it will be turned on exactly once, or exactly twice and so on. Over a sufficiently long period of time though, the event will occur on average once a week. Such events follow what is known as a Poisson distribution which is given by the standard equation:

$$P = \frac{\mu^n}{n!} e^{-\mu}$$

where (in this context):

P_n = probability of exactly n switches per unit time

μ = average number of switches per unit time

In the example calculation in CIRIA publication R152 μ was once per week or about 50 times per year. Using the equation above the probability of the light <u>not</u> being switched on at all within the week (ie n = 0) is 0.37, being switched on exactly once (ie n = 1) is 0.37, (as μ has to be 1 per unit time) <u>exactly</u> twice (ie n = 2) is 0.18 and so on. Accordingly, the probability that the light will be turned on <u>at least</u> once in a given week is 0.63 (ie 1 – 0.37).

Considering the event over a period of one year, the probability that it will not be turned on for a whole year is, unsurprisingly, negligible (less than 1 in 10^{22}). The probability that the light will be turned on <u>at least</u> once in a year is effectively 1.0.

Following on from this the probability of an explosion is given by:

Annual probability of explosion = annual probability of ignition × annual probability of protective system failure

A5.2.1 Factors to consider in a fault tree analysis for soil gas

To undertake quantitative risk analysis for gas, mathematical models are required to represent the following:

- variation of gas concentrations and flow rates

- gas generation
- lateral gas migration
- gas surface emissions and entry into buildings
- gas seepage rates into trenches or other excavations
- ventilation of buildings and underfloor voids
- air flow rates and pressures in granular materials or void formers for positive air systems.

A5.2.2 Sensitivity analysis

A sensitivity analysis should be undertaken for all quantitative risk assessments to identify the risks associated with slight variations of input parameters. For example the sensitivity of the assessment to variations in gas concentrations can be assessed (see Figure A5.3):

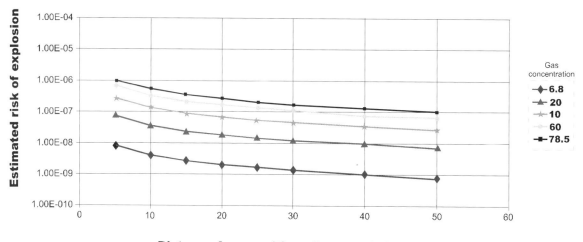

Figure A5.3 *Example of sensitivity analysis*

A useful method of justifying the input parameters and also defining the likely sensitivity of each one is summarised in Table A5.2.

Table A5.2   *Example of baseline and sensitivity parameters*

Parameter	Baseline value	Range for sensitivity analysis	Justification
Methane concentration	6.7 % v/v	6.7 % v/v to 78.5 % v/v	To cause migration, concentration should be consistent over period of migration. Gas levels are variable and using maximum values is unrealistic as they do not occur all over site or all the time. So use average values and check with maximum values in sensitivity analysis.
Carbon dioxide concentration and oxygen depletion	N/A	N/A	The critical level of methane is 1 % v/v. The critical level for carbon dioxide is 1.5 % v/v and for oxygen depletion any gas must remove oxygen so that levels fall to 17 % v/v or below before effects take place (CIRIA publication R149). This requires more than 1 % v/v of gas. So worse case is 1 % v/v design criteria.
Borehole flow rate	13.6 l/h	N/A	Borehole flow rate is maximum recorded in the site and therefore no sensitivity analysis has been used for this parameter.
Gas pressure	20 Pa	N/A	Value is very worse case based on experience of monitoring pressures in similar types of material.
Atmospheric pressure drop	2900 Pa	N/A	Worse case value that occurred when Loscoe explosion occurred (where there is 29 milibar, which is equivalent to 2900Pa, drop in seven hours). See Section 2.6.3.
Thickness of migration pathway	4 m	N/A	Worse case maximum thickness of made ground below site.
Permeability of soil to water (hydraulic conductivity)	1×10^{-5} m/s	1×10^{-3} m/s to 1×10^{-7} m/s	Soils described as clayey silty sands. The permeability will be governed by the fines content. The permeability with a worse case for fine sand or silt is usually 1×10^{-5} m/s, but could be as low as 1×10^{-7} m/s. This will be converted to intrinsic permeability for analysis of gas flow.
Diffusion coefficient for soil	1×10^{-6} m²/s	N/A	This value is from CIRIA publication R152 for a soil with a voids ratio of 0.25. The soils in this site are unlikely to have a voids ratio greater than 0.3 and the table in CIRIA publication R152 indicates this will not significantly affect the diffusion coefficient.
Viscosity of soil gas	1.03×10^{-5} Ns/m²	N/A	From CIRIA publication R152 value for methane.
Density of soil gas	0.717 kg/m³	N/A	From CIRIA publication R152 value for methane.
Ventilation of confined space in building	2 Air changes per day	N/A	Worse case value based on CIRIA publication R152 and Sladen *et al*.
Defects in gas membranes	17 defects per ha	30 defects per Ha	Membranes were installed by specialist sub-contractor, under QA system and independently verified.
Size of house	6 m × 8 m	N/A	Typical size of individual unit from layout plans. Longer terraces will have negligible effect on calculations.
Size of cupboard	1 m × 1 m × 2 m	N/A	Cupboard with light switch or other electrical source of ignition. From floor plans of development.

Note:

The example above presents sensitivity analysis ranges and baseline data for a number of parameters. There are a number of other parameters that should also be considered for sensitivity analysis.

N/A = Not available

A5.2.3   Limitations of fault tree analysis

Fault tree analysis is highly effective in determining how combinations of events and failures can cause specific system failures; however the technique has three main limitations:

1. Fault tree analysis examines only one specific event of interest and so has a narrow focus. To analyse other events, other fault trees must be developed.

2. Fault tree analysis is as much an art as a science and a significant degree of judgement is required. The level of detail, types of events included and organisation of the tree vary significantly between analysts. However, given the same scope of analysis and limiting assumptions, different analysts should produce comparable, although not identical, results.

3. Using fault tree analysis results to make statistical predictions about future system performance is complex and requires a great deal of experience.

In addition, the focus of many risk assessments often becomes concentrated on equipment and systems, and human and organisational issues are not addressed adequately. For example, in gas risk assessments how can the risk of accidental membrane damage or blocking of vents by residents in a housing development be accurately predicted?

Fault tree analysis is unlikely to be necessary except on the most difficult of gassing sites and for the majority of situations qualitative or semi-qualitative methods of risk analysis will be sufficient. It is most appropriate for sites where the risk is considered to be close to the tolerable limit or where the remediation costs could potentially be unacceptably high.

A5.3 VARIATION OF GAS CONCENTRATIONS AND FLOW RATES

It has been usual practice, in most cases, to use worse case values of gas concentrations and flow rates in risk assessments, probably due to the limited gas data that is available on many sites.

If there is a sufficient data set, statistical analysis of the results can be undertaken to determine the mean value or the results can be plotted as a frequency distribution to determine the modal value. Normally the frequency distribution is based on ranges of gas values as shown in Figure A5.4.

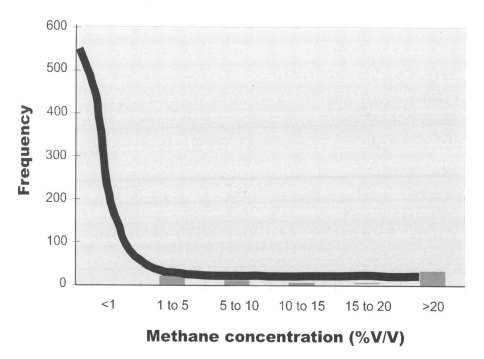

Figure A5.4 *Histogram of monitoring results*

The frequency distribution can also be used to estimate the type of probability distribution that applies to a data set. For a continuous function, the probability density function (pdf) is the probability that the variate has the value x; thus the probability of a parameter exceeding a particular value can be estimated. For example the probability that gas emissions overwhelm an underfloor venting system can be estimated.

From Figure A5.4 above:

$$f(x) = \frac{1}{\mu} e^{-\frac{x}{\mu}}$$

and

$$P(x < x_0) = 1 - e^{-\frac{x_0}{\mu}} \quad \text{Where } \mu = \text{mean of data}$$

So the probability that the methane concentration is less than one per cent from the above graph (mean of data is two per cent) is given by

P(<1 %) = 1 − e$^{−1/2}$ = 0.39

A similar process can be undertaken for different shaped distributions.

A5.3.1 Simple gas generation model

The latest guidance on the management of gas generated from landfill sites is published by the Environment Agency (2004). This states that:

"An initial approximation of the landfill gas generation for sites that have taken biodegradable waste can be produced by simply assuming that each tonne of biodegradable waste will produce 10 m³ of methane per year. The following calculation will then give an approximation of the gas flows that would be generated. This equation will produce an overestimate of gas flow at peak production and gas flow from historical waste deposits. Alternative and more sophisticated models of gas generation (including GasSim) can be used at this stage should the operator wish to."

Q = M × 10 × T / 8760

where:

Q = methane flow in m³/hour

M = annual quantity of biodegradable waste in tonnes

T = time in years that waste was placed.

GaSSim or GASSIM Lite uses more complex equations based on a number of factors including the quantity and composition of the waste, the infiltration of rainfall, moisture conditions and cellulose decay rates. The programme can be used to give a gas generation over the life of a landfill in appropriate cases (Figure A5.5).

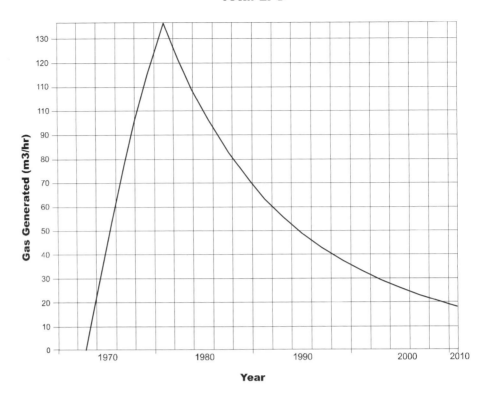

Figure A5.5 *Landfill gas generation profile from GasSim Lite*

GasSim is discussed in more detail later in this appendix.

A5.3.2 Simple lateral gas migration models

Gas surface emissions and entry into buildings

The prediction of gas surface emissions is complex and there will be a great deal of uncertainty associated with this critical input parameter. The most common way to predict surface emission rates is by using the method proposed by Pecksen (1986). This is based on a simple relationship between borehole flow rate and radius of influence of the borehole. In the example provided by Pecksen a radius of influence is assumed and this gives an area of influence if 10 m² (ie the gas from the borehole would be emitted over a 10 m² area if the borehole was not present, equivalent to a radius of influence of 1.78 m).

Potential surface emission rate (l/m²/h) = borehole gas emission rate (l/h)/10

The main disadvantage of this method is that it assumes general vertical migration when gas tends to escape via discrete fissures. Gas surface emission rates can also be estimated using the results of the generation calculations and compared to measured flow rates on a site. Again the method gives only a crude approximation of emission rates.

For example, based on the generation curve in Figure A5.5, in 2009 the gas generation is estimated to be 20 m³/h. If this was estimated from a site that is 500 m by 200 m in area and assuming all the gas is emitted through the surface as methane (a very conservative assumption) then the surface emission rate will be:

(20 m³ × 1000)/(500 m × 200 m) = 0.2 l/h

Gas surface emission rates can also be confirmed using flux chambers where appropriate. However flux chambers should never be used as the only method to estimate emissions. Neither should they be used to estimate emissions occurring from lateral migration, or where the gassing source is at depth and there is a risk it could migrate by discrete vertical pathways (unless the flux chamber testing is carried out over all identified pathways).

A5.3.3 Simple pressure drop generation model

A simple calculation can be carried out that takes into account worst case assumptions (such as falls in atmospheric pressure) when calculating gas ingress into a building. This calculation may be used, in addition to the monitoring data, laboratory analysis and qualitative risk assessment, when considering the need for remedial or protective measures.

Example:

A site investigation and subsequent gas monitoring has identified elevated methane concentrations in the vicinity of a small house:

- maximum methane concentration is 2.7 per cent
- depth of unsaturated rock under the house is 3 m
- area of floor of house is approximately 64 m²
- porosity of ground is 46 per cent (fractured granite based on physical and chemical hydrogeology, Domenico and Schwartz, 1997).

The volume of gas at source underlying the house (V_1) is:

$$V_1 = 64 \times 3 \times 46\%$$
$$V_1 = 88.3 \text{ m}^3$$

Assuming that gas moves in response to a fall in atmospheric pressure (assume 10 mbar in three hours), and that temperature remains constant, the volume of gas will increase as follows:

$$P_1 V_1 = P_2 V_2$$
$$1010 \times 88.3 = 1000 \times V_2$$
$$V_2 = 1010 \times 6.9 = 89.2 \text{ m}^2$$

"Additional" gas generated by this pressure fall is calculated as V_3:

$$V_3 = V_2 - V_1 = 0.9 \text{ m}^3$$

For an absolute worst case calculation, it is assumed that, as the site is small, all soil gas under the house, enters and accumulates in a small space (eg toilet). It is assumed that no dilution occurs as the gas migrates and that there is no ventilation within the toilet.

Typical volume of toilet:

$$2.0 \times 1.0 \times 1.5 = 3.0 \text{ m}^3$$

Volume of methane entering the toilet:

$$V_3 \times \text{concentration of methane} = 0.9 \times 0.027 = 0.02 \text{ m}^3$$

Concentration of methane in the toilet:

Volume of methane entering toilet/volume of toilet × 100 = 0.02 / 3 × 100 = 0.7 %

This concentration is an order of magnitude below the LEL for methane of 5 per cent by volume.

A5.4 LANDFILL GAS GENERATION AND MIGRATION MODELS

Many models are available worldwide for the calculation of gas generation from anaerobic degradation of wastes with the models usually being designed for landfill gas generation within landfills (the main aspects of which are shown in Figure A5.6). Examples of these models include GasSim (UK), TNO (Netherlands), Afvalzorg (Netherlands) and EPER Model Ademe (France). The majority of these landfill gas models are based on first order decay modelling techniques; some also employ multi-phase calculations. Other models being used outside of the EU include the US EPA LandGEM model which has been adopted and modified by other countries including Australia which has developed its own LABS spreadsheet based on the same methodology.

The principal drivers behind the development of HELGA (Gregory *et al*, 1999) and subsequently GasSim, were the concerns of the potential health effects of living near and working on landfills, and the need to substantiate these, as well as the need for a management tool to help the UK meet international agreements to reduce the emissions of greenhouse gases to the environment. GasSim uses the modelling principles that were developed under the HELGA framework collating them in a user-friendly MS Windows-driven software package (although experience of landfill gas and landfill engineering is required).

GasSim has been designed to:

- assess the risks from current and planned landfill gas emissions
- provide a framework that will contribute to the assessment and validation of the inventory of burdens associated with the landfilling of wastes
- help regulators and other relevant organisations compare the relative risks associated with different landfill gas management techniques.

GasSim consists of the following modules to aid in risk assessment:

- source term
- emissions model
- environmental transport
- exposure/impact.

GasSim is a probabilistic model based on Monte Carlo Simulation using probability distribution functions for the majority of model inputs. This allows uncertainty to be calculated based on site inputs. The interaction of the different modules is then expressed in the conceptual model for a particular landfill.

Of the functions and model outputs available from GasSim, the following are of interest for risk assessment from soil gases:

- landfill gas generation calculation including typical concentrations of trace constituents in landfill gas

- calculation of surface flux
- lateral migration
- vegetation stress
- exposure.

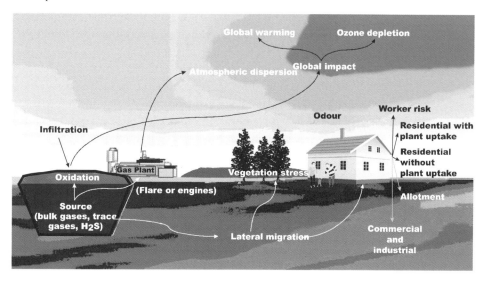

Figure A5.6 *Typical conceptual model*

A5.4.1 Source term

The GasSim source term module determines the generation of landfill gas for an individual landfill site based on the mass of waste deposited and the waste composition of the waste stream or streams. The model calculates the gas generation from the wastes based on a multi-phase first order decay model further details of which can be obtained from the user manual (Environment Agency, 2002). The calculation adopts the following equation:

$$C_t = C_0 - (C_{0,1}e^{(-k1t)} + C_{0,2}e^{(-k2t)} + C_{0,3}e^{(-k3t)})$$

and

$$C_x = C_t - C_{t-1}$$

where:

C_t = mass of degradable carbon degraded up to time t (tonnes)

C_0 = mass of degradable carbon at time t = 0 (tonnes)

$C_{0,i}$ = mass of degradable carbon at time t = 0 in each fraction (1, 2, 3 rapidly, moderately and slowly degradable fractions respectively) (tonnes)

C_x = mass of carbon degraded in year t (tonnes)

t = time between waste emplacement and LFG generation (yrs)

ki = degradation rate constant for each fraction of degradable carbon (per yr)

Waste input data are entered in tonnes per year. The types of wastes filled can then be chosen from default waste types such as "domestic" or "industrial" that have composition analyses based on HELGA (Gregory *et al*, 1999). If the composition of the waste is known then user-defined waste types can be generated based on its constituent parts such as paper, card, inserts etc. The waste type can also be entered based on site-specific data using its fraction of cellulose, hemi-cellulose, water content and an estimation of its decomposition potential to allow the model to calculate available carbon for degradation.

Utilising the waste data and other model inputs such as waste moisture content the model is able to calculate the gas generation from a landfill in m^3/h of landfill gas. The proportion of methane to carbon dioxide within the landfill gas can be user-defined.

The model simulates the generation of trace gases within the generated landfill gas by assuming certain trace gas concentrations within the generated gas based on data gathered from landfills. The model also assumes declining trace gas emissions based on a trace gas half life derived from many different research studies conducted on a number of landfills. The trace gas concentrations can either be selected from five sets of default values including a set derived from a number of research studies (AERC draft database, 2001; Derwent *et al*, 1996; and Stoddart *et al*, 1999); a set for pollution inventory reporting; a set for odorous compounds; and two data sets for US co-disposal and non co-disposal based on the US EPA LandGEM model which itself is based on AP-42 and CAA values.

The model can also simulate hydrogen sulphide production within the source term. The model simulates hydrogen sulphide production pseudo-mechanistically, using simulated microbial degradation pathways. Additional model inputs are required to assess hydrogen sulphide generation. Default inputs are provided however the user manual strongly recommends that site-specific values are used. The additional inputs include leachability of iron, leachability of sulphate, iron in leachate, calcium sulphate in leachate, half life of calcium sulphate, and also an assessment of calcium sulphate and iron in the waste streams (%).

The emission model calculates the landfill gas emission of bulk and trace gases to the environment after allowing for landfill gas collection, flaring, utilisation (energy recovery), and biological methane oxidation. It is assumed that landfill gas generated and not collected is in equilibrium and will be emitted from the landfill cap or liner at a steady state. Additionally, the model calculates the concentrations of other major and trace gases emitted from the landfill surface, landfill sides, flares and engines.

The source term module may have some limited value in assessing risks from hazardous soil gases generated from waste in historical closed landfills by back-analysing the residual potential for buried wastes to produce landfill gas. However, this would need to be based on a detailed knowledge of the waste from either the historical data available on the type, or the quantity, moisture content and age of the wastes. These data are not often likely to be available for older landfills and contaminated sites.

The estimation of gas generation from all landfill gas models is often inaccurate due to the variability in site conditions and waste. The accuracy of various landfill gas generation models including GasSim is discussed in a paper by Jacobs and Scharff (1990). The paper concludes that site verification of gas generation should be conducted to allow model calibration, for example long-term gas pumping trials from landfills.

The determination of trace gases and hydrogen sulphide from gassing ground would best be assessed from site-specific measures rather than adopting GasSim defaults that are based on data gathered from landfills. The model will allow user-defined inputs for trace gas concentrations. Trace gas generation concentrations will then pass into the lateral migration module.

A5.4.2 Surface and lateral emissions in GasSim

Surface and lateral emissions from the model are calculated from the residual gas within the site, that is the quantity of gas that remains within the site after active landfill gas extraction. The quantity of emission through both the cap and liner are determined by the permeability and thickness of the most impervious layer. The model assumes that gas movement is via plug flow with both the cap and liner being homogenous in nature. The emissions are then calculated using Darcy's Law for homogenous media. This equation has been modified from HELGA by including a landfill surface area input.

Emissions from uncapped areas of the landfill are assumed to occur through the surface of the landfill as this will offer the route of least resistance. Also, lateral migration of residual gas is considered to occur only through the unsaturated zone of the landfill.

Surface emissions

Surface emissions are calculated from a simple model that considers methane oxidation within the landfill capping layer and also gas loss through fissures. Both of these inputs are available as defaults with the methane oxidation being based on the Intergovernmental Panel on Climate Change (IPCC, 1996) or user-specified values can be used. The model assumes that methane that is oxidised will produce carbon dioxide and adds this to the surface emissions.

The model will be useful only for estimation of methane and carbon dioxide flux from ground containing wastes or situations where the source term emission is known. The accuracy of the model would be improved if based on site-specific measurements as user defined inputs. However, an input for methane oxidation rate is difficult to determine.

Lateral migration

GasSim adopts a simple approach to modelling lateral gas migration that promotes conservative estimates of migration for risk assessment. The model presumes that the gas migrates horizontally through the largest pores or fissures assuming a one-dimensional linear pathway. The model does not consider buoyancy-driven flow and temperature-driven flow and is based on advection and diffusion. The model also assumes that no biological oxidation, dispersion, retardation or other attenuation/ reaction processes occur that will reduce the concentration of the gas.

Typically, for PPC applications, the GasSim output will be supplemented with further risk assessment risk matrices according to the DETR and EA guidance <www.gassim.co.uk>. The model is also limited as it considers only a minimum time step of one year.

A5.4.3 Vegetation stress

Determination of vegetation stress is based on the lateral migration module. The limit of vegetation stress is based on research which suggests that vegetation stress can be caused by concentrations of 5–10 % v/v of carbon dioxide and approximately 45 % v/v methane in the root zone. As methane can be oxidised to carbon dioxide the model calculates the sum of the methane and carbon dioxide concentrations which is then compared to the assumed vegetation stress threshold.

The vegetation stress model is simplistic and subject to the limitations of the lateral migration model.

A5.4.4 Exposure

The exposure module is used to estimate the risk to off-site receptors from the intake of pollutants arising from a landfill and its associated plant. The model considers the characteristics and behaviour of the people who are most likely to be affected by the pollutants, and calculates exposure concentrations in the relevant media to which the people are exposed. The likely total dose of each contaminant from the exposure pathways in the conceptual model set up can be predicted for a choice of exposure periods. The intake models for inhalation, skin exposure and ingestion are based on the CLEA method as described in CLR10 (Defra and Environment Agency, 2002d).

In addition, GasSim also allows evaluation of the inhalation exposure of site workers. For screening purposes, GasSim calculates the air concentrations of pollutants generated by landfill surface emissions, flare emissions and gas engine emissions. These are a direct output from the atmospheric dispersion module, and a minimum dispersion distance of 1 m is simulated. The atmospheric dispersion model is based currently on a Gaussian Plume Model (Draxler, 1981). A new version of GasSim is due to be released shortly that will include more complex new generation air dispersion modelling.

A5.4.5 GasSim context for assessing risks posed by hazardous soil gases to buildings

GasSim has been designed specifically for use by landfill operators to address risk assessment required for Pollution Prevention Control (PPC) applications with a focus on landfill management options. The package as a whole is likely to have limited value for assessment of risk for hazardous soil gases at new developments. The model has been designed to fit into the conceptual approach to risk assessment as described by the DETR and EA (2000) and, in general terms, addresses Tier 2 with some modules addressing Tier 3. Limitations in the lateral migration model often justify the addition of risk matrices (also based on the DETR guidelines) to a PPC application for the assessment of risk posed to off-site receptors.

The model depends on accurate determination of the source term inputs based on waste types, data which in general will be available only from modern landfilling practices. The model has a minimum time step of one year so cannot simulate short-term risk (for example, lateral migration associated with sudden drops in atmospheric pressure or acute exposure resulting in asphyxiation or other acute health effects).

For contaminated ground containing buried biodegradable wastes there is likely to be insufficient information to suit the required source term input parameters required for the model. In these cases it will be more accurate to assess risk based on observed gas generation rates and concentrations, or to use a contaminated land software package that estimates partitioning into the vapour phase from concentrations of contaminants in soil or groundwater. In some cases, gathered data may be used to calibrate a gas generation curve, but that this would add little benefit to the risk assessment process.

For assessment of risk for developments next to landfills the same limitations will apply, that is if a hazard for a development is posed by an older landfill there may not be the historical data available from the site to accurately assess the source term. The model can be calibrated through site investigations aimed to address the source term inputs and gas generation rate (particularly if gas extraction is installed at the landfill) which may then be used for risk assessment. However, the model will be limited by the simplicity of the lateral migration module.

A5.5 MATHEMATICAL MODELS AND ALLOWANCE FOR GAS PROTECTION IN RISK ASSESSMENT

The following factors should be considered when allowing for gas protective systems in the risk assessment process. Additional factors have been summarised in Table A5.3.

Passive ventilation is preferred as it requires less maintenance than systems with pumps. This is especially so in residential developments. Where active systems are used they should be able to perform passively in the event of fan failure. The level of ventilation can be calculated using well-proven methods described in CIRIA publication R149 that have been demonstrated to provide an adequate indication of likely performance (Pecksen, 1986 and Wilson *et al*, 1999). The level of risk can be estimated, based on the level of ventilation provided and the factors of safety built into the calculations.

Where active venting is installed measures should be put in place to ensure maintenance is carried out and the robustness of these measures can be allowed for in the risk assessment. This applies equally to dilution or proprietary positive air blanket systems. There are numerous examples of all types of active systems not being maintained in the past and any mechanical system can fail.

The type of floor construction has a significant effect on the likelihood of gas ingress into a building. Reinforced concrete, cast *in-situ* suspended slabs that are well detailed, sealed joints, and cast in service penetrations should be relatively gas resistant.

The type of building will affect the risk of gas build up. Open buildings, such as warehouses with no confined spaces for gas ingress, are not particularly sensitive. However it is important to consider areas such as offices, mess rooms etc, and locations where services enter these accommodation areas, as these are more sensitive in terms of gas risk.

One of main causes of gas ingress into buildings is not sealing service ducts, and gas gets into them and enters the building. These should be sealed with expanding foam where they enter buildings.

All proprietary gas membranes have varying advantages and disadvantages. However, all of them have sufficient gas resistance in the material itself for the vast majority of gases and vapours that are encountered on redevelopment sites. Surviving installation intact is the key issue with membranes. For example 1200 g DPM material is unlikely to be installed intact. A list of factors that affect the risk of membrane damage is provided in Box A5.1.

Box A5.1 *Estimating membrane damage*

The probability of damage occurring to a membrane can be estimated based on data produced by the Environment Agency for landfill liner systems (Environment Agency, 2004d). Using this guidance the base probability density function for the number of defects per hectare of a membrane liner during installation is summarised below for good quality control.

Estimated hole frequencies per hectare of liner

Category	Pinholes (0.1–5 mm^2)			Holes (5–100 mm^2)			Tears (100–10 000 mm^2)			Total		
	Minimum	Most likely	Maximum	Minimum	Most likely	Maximum	Minimum	Most likely	Maximum	Minimum	Most likely	Maximum
Good case	0	10	15	0	5	10	0	2	5	0	17	30

This base probability can then be adjusted up or down, depending upon the site, and material specific factors that affect the probability of membrane damage on any development.

Factor	Adjustment to probability
Design life – as design life increases the risk of damage occurring increases.	Increase probability of failure by 10 per cent for every 20 years of design life over 50 years
CQA – if a membrane is installed under a CQA system and joints are tested, the risk of defects is reduced. Conversely without a CQA system the risk of installation defects increases.	If membrane installed without CQA and seam testing, increase probability of defects by 300 per cent. If membrane is installed without seam testing but with CQA, increase probability of defects by 150 %.
Specialist installers – if specialist installers are used the risk of construction defects is reduced and, conversely, if the membrane is installed by ground workers, brick layers or other similar staff the risk of defects is increased.	If membrane is not installed by specialists increase probability of failure by 100 %.
Independent inspection – if the membrane is inspected by an independent consultant immediately before screeding or concreting, the risk of construction defects is reduced.	If there is no independent inspection immediately before covering the membrane, increase the probability of defects by 200 %.
Area of membrane - smaller areas have a proportionately greater number of seams. Landfill liner data is based on large areas of installation and most developments will require smaller areas.	Increase probability of failure by 200 % for domestic installations and 100 % for large commercial or industrial installations
Complex details – number of service penetrations or complex structural slab and foundation details.	If there are complex foundation details or a large number of service penetrations that a membrane should be sealed to, then increase probability of failure by 300 %.
Type of membrane – thin damp proof membrane (DPM) materials will have a greater risk of defects than thicker and more robust specific gas resistant membranes. As the strength and puncture resistance decrease the probability of defects increases.	For thin damp-proof membrane (DPM) material increase probability of failure defects 1200 g – 400 % 1000 g – 600 %.
Protective layer – if a protective layer such as a geotextile fleece or sand layer is not provided the risk of defects increases.	If no protective layer is provided increase probability of failure by 200 %.
Settlement – the likely differential settlement between building components will affect the risk of membrane failure. The greater the level of likely movement, the greater the risk.	Site-specific assessment based on likely level of movement
Position in construction – membranes placed below a concrete slab face less risk of accidental damage by the occupiers than those above the slab and below a screed.	If membrane is above structural floor slab increase probability of damage by 300 %.
Development – the risk of accidental membrane damage is greater in houses than in large commercial buildings.	For houses increase risk of accidental damage by 300 %.

The suitability of any system for a particular gas regime should be demonstrated by calculation or substantial monitoring (preferably continuous) over a suitable period. The disadvantage of relying on monitoring to demonstrate performance is that temporal changes in gas generation and emission cannot be allowed for. The flow of air or gas through the ground granular layers and likely air flow routes is particularly important. For dilution systems the air flow should be diluted to safe levels. For positive air flow systems it should be clearly demonstrated that the air flow is sufficient to counteract the gas flow from the ground and that short circuits in the system will not occur (which in effect can turn it into a venting/dilution system). To allow representative comparisons to be made, gas monitoring in wells will be required at the same time as monitoring of the gas dilution within the system.

A5.6 VENTILATION

To undertake a quantitative risk assessment, the ventilation capacity of underfloor venting systems, positive air flow systems and indoor spaces has to be estimated by calculation.

The most commonly used method for passive systems is that provided in British Standard BS 5925:1991 *Code of Practice for ventilation principles and designing for natural ventilation*. This has been well-proven by performance monitoring of various underfloor gas protective systems. Further information is provided in Box A5.3.

Box A5.2 *Ventilation design*

> Determine the ventilation required for a housing development up to three storeys high located on the Southampton coast. The height of the vents is 0.15 m and the height of the building is 7 m. The required flow of fresh air through the void is 1.8 m³/h for a building that is 40 m long and 20 m wide. The underfloor void is 150 mm high and the surface emission rate is $2.3 \times 10\text{-}5 \text{m}^3/\text{h}/\text{m}^2$
>
> From Figure 5 in BS 5925: 1991 *Code of practice for ventilation principles and designing for natural ventilation*. Hourly mean wind speed, U_{50} 4.5 m/s (measured at 10 m height in open terrain)
>
> Determine correction ratio from Table 9 of BS 5925. Allow for the design wind speed being exceeded 80 per cent of the time (ie this is the worst case value and gives the highest confidence that the passive system will operate), and consider an exposed coastal location So factor = 0.56
>
> $U_m = U_{50} \times 0.56 = 4.5 \times 0.56 = 2.52$ m/s
>
> Determine factor K from Table 8 of BS 5925. Assume an urban environment so $K = 0.35$. These factors amend the mean hourly wind speed to allow for differing terrain and different heights. The pressure on the side of the building is governed by the height of the building but, to be conservative, we will use the height of the vent as the design height.
>
> So reference wind speed $u_r = u_m * K * z^a = 2.52 \times 0.35 \times 0.15^{0.25} = 0.55$ m/s from BS 5925:1991.
>
> Now calculate required vent A_w area to give flow of fresh air, Q. Assume the discharge coefficient for a narrow opening, $C_d = 0.61$, which is a typical value for narrow openings from BS 5925:1991. (This is a factor that correlates theoretical performance to actual performance)
>
> We do not know the orientation of buildings so use pessimistic value of ΔC_p
>
> $\Delta C_p = 0.4$
>
> Area of ventilation required for whole building, A_w calculated using the following equation from BS 5925: 1991
>
> $Q = 1.8$ m³/h and we need to convert this to m³/s = $1.8/3600 = 0.0005$ m³/s
>
> $$A_w = \frac{Q}{U_r \times C_d \times \sqrt{\Delta C_p}} \times 10^6$$
>
> $$= \frac{0.0005}{0.55 \times 0.61 \times \sqrt{0.4}} \times 10^6 = \underline{\mathbf{2360 \text{ mm}^2}}$$
>
> This is equal to 2360 mm²/40 m = 60 mm²/m. This is less than the minimum venting area required in the Building Regulations of 1500 mm²/m, so we need to provide the minimum vent area. This can be achieved by using normal air bricks with a vent area of 6000 mm² at 4 m centres.

Box A5.3 *Ventilation design (contd)*

> Check time to fill the void to five per cent methane if there is no wind
>
> Time to fill to five per cent methane = (Volume of void × 5 %)/surface emission rate of gas below building
>
> The plan area is 40 m by 20 m and volume of void is 40 × 20 × 0.15 = 120 m³
>
> Time to fill = (120 m³ × 0.05)/(2.3 × 10⁻⁵ m³/h/m² × 40 m × 20 m) = 326 hours
>
> This is less than the maximum period of still wind of 10 hours reported in the Partners in Technology report (DETR, 1997).
>
> **Note:** It has not been normal practice in ventilated void design to increase the vent area to allow for reduced effective vents in series described in BS 5925 which would increase A_w by a factor of 1.4 in most cases.
>
> This is because the calculations are widely accepted as already being conservative (for example ignoring temperature effects using low values of ΔC_p and using onerous values of gas emission rates). Also in the vast majority of cases, the overriding requirement is to produce the minimum ventilation requirements of 1500 mm²/m of wall, as in this case.

The performance assessment of the ventilation layer below a building should take account of site-specific factors and the sensitivity of the building. For example, a lower standard of performance may be acceptable for a warehouse than for housing. The ventilation performance of an underfloor venting system should be defined in accordance with the grading system used in the Partners in Technology report.

Where void formers are used below larger floor slabs the manufacturers normally provide details of testing of modelling that indicate air flow rates. The permeability of granular materials through which air flow needs to be modelled can be obtained from any standard soil mechanics reference book.

VERY GOOD	The steady state concentration of methane over 100 per cent of the ventilation layer remains below one per cent by volume at a wind speed of 0.3 m/s.
GOOD	The steady state concentration of methane over 100 per cent of the ventilation layer remains below one per cent by volume at a wind speed of 1 m/s and below 2.5 per cent by volume (50 per cent LEL) at a wind speed of 0.3 m/s.
FAIR	The steady state concentration of methane over 100 per cent of the ventilation layer remains below one per cent by volume at a wind speed of 2 m/s and below five per cent by volume (100 per cent LEL) at wind speeds of 0.3 m/s.
POOR	The steady state concentration of methane over 100 per cent of the ventilation layer remains below one per cent by volume at a wind speed of 3 m/s and below five per cent by volume (100 per cent LEL) at a wind speed of 1 m/s.
UNSUITABLE	The steady state concentration of methane over 100 per cent of the ventilation layer remains below one per cent by volume at a wind speed of 3 m/s and above five per cent by volume (100 per cent LEL) at a wind speed of 1 m/s.

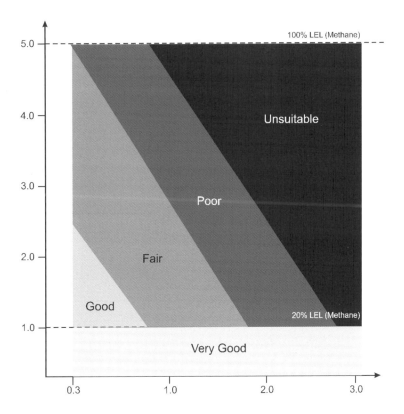

Performance assessment criteria of ventilation layer for methane hazard

The probability of failure of the venting can then be estimated by defining the required gas emissions to exceed venting capacity and calculate probability of this occurring. However other scenarios that should be looked at, but are more difficult to apply probabilities to, include the risk of vents being blocked, the risk of still air conditions, etc.

For active systems the fan air flow rates can be specified and similar calculations can be undertaken. The manufacturers of fan systems should also be able to provide performance data on the reliability of the fans.

A5.7 OTHER PROTECTIVE MEASURES

The probability of failure of other parts of the gas protective system also needs to be assessed. For example gas membranes, venting trenches, gas detection and monitoring systems.

A5.7.1 Frequency of exposure

The frequency of exposure depends on site-specific factors and possible events that could lead to exposure are:

- methane explosion – light switches, central heating switches, burglar alarms, electrical appliances, etc
- asphyxiation – presence in excavation or room with a gas cloud.

Table A5.3 *Factors affecting risk on gassing sites*

	Source variables			
Increasing risk ↓	**Location of source** Off site On site adjacent to building On site beneath building but below non gassing soils Immediately below building floor slab	**Generation potential of source** Negligible Very low Low Moderate High Very high	**Risk rating of gas regime** Low concentration/borehole flow less than limit of detection Low concentration/low borehole flow rate Low concentration/high borehole flow rate High concentration/low borehole flow rate High concentration/high borehole flow rate	Increasing risk ↓
Increasing risk ↓	**Area of gassing material** Very small Small area Medium area Large area Very large area	**Thickness of gassing material** Very thin (less than 0.3 m) Thin (less than 1 m) Medium (up to 5 m) Thick (up to 10 m) Very thick (greater than 10 m)		Increasing risk ↓
	Pathway variables			
Increasing risk ↓	**Pathway permeability** Cohesive soils Mixture of cohesive and granular soils high water table Mixture of cohesive and granular soils water table at depth Predominantly granular soils/fractured rock, high water table Predominantly granular soils/fractured rock, water table at depth Presence of discrete pathways such as underground services with permeable surround, faults, natural or man-made cavities		**Foundation conditions** Reinforced concrete raft Shallow footings Bored, cast *in-situ* piles Driven piles Stone columns	Increasing risk ↓
	Target variables			
Increasing risk ↓	**Complexity of substructure** Suspended floor slab with flat underside Single level ground bearing-floor slab with flat underside Single level ground bearing-floor slab with monolithic downstanding ground beams Multi-level ground bearing-floor slab Basement incorporated in structure	**Installation of protective measures** Specialist contractor installed, independent CQA inspection and testing by specialist engineer Specialist contractor installed with CQA, independent inspection by specialist engineer Specialist contractor installed, no CQA and no independent inspection Main contractor installed, independent inspection by specialist engineer Main contractor installed DPM, no independent inspection	**Maintenance requirements** Simple routine inspection (most passive systems) Routine electrical and mechanical maintenance (most active systems) Complex electrical and mechanical maintenance (rare)	Increasing risk ↓
Increasing risk ↓	**Sensitivity of end use** Hardstanding/heavy industrial Any use with basement car park Light industrial Offices/shops Hospitals, care institutions, schools Domestic housing			Increasing risk ↓
Increasing effectiveness ↓	**Effectiveness of underfloor venting media** Well graded gravel (eg Type 1) Pipes or geocomposites in well graded gravel Open graded gravel (eg 4/40 or 4/20) Pipes or geocomposites in open graded gravel Thin void formers (up to 20 mm thick, polystyrene, geocomposites, etc) Thick void formers (over 20 mm thick, polystyrene, geocomposites, etc) Open void	**Effectiveness of membrane barriers** 1200 g DPM (note this is unlikely to survive construction intact) 2000 g unreinforced DPM Less than 1mm thick membranes Asphalt/bitumen barriers Specific gas resistant membranes 1mm or thicker (eg HDPE, polypropylene, HDPE/aluminium foil sandwich)		Increasing effectiveness ↓

Explanatory notes to Table A5.3

Sensitivity of end use	Issues that affect sensitivity of end use are occupancy rates, rate and nature of internal ventilation, management systems that are in place, public perception
Location of source	The closer to the source, the greater the risk of gas being able to enter a building
Generation potential of source	Generation potential is affected by age of gassing material, degradable content, nature of degradable content (finely chopped material is more easily degraded than large pieces), moisture content (in the past and in future), fluctuation in water table (especially tidal areas)
Area and thickness of gassing material	The volume of gassing material affects the volumes of gas that can be generated. Continuous generation of relatively large volumes is required to maintain gas migration from the ground.
Pathway permeability	Pathway permeability is affected by soil types and by the presence of discrete discontinuities such as services, tunnels, mineworkings, water table etc
Risk rating of gas regime	Risk rating of gas regime is affected by soil gas concentrations and borehole flow rates. High borehole flows are considered more of a risk than high concentrations
Foundation conditions	Foundations such as piles and stone columns can provide pathways for gas to reach the underside of the building. Some ground stabilisation processes can reduce soil permeability below the slab
Complexity of substructure	The complexity of the underside of slabs affects air flow for all types of underfloor venting or proprietary active pressurisation/exhaust systems. As conditions become more complex there is an increased risk of dead spots, short circuits and reduced flow in both passive venting and proprietary active venting systems due to constrictions, loss of air pressure from proprietary positive systems, etc. Detailing membranes becomes more difficult and the stresses imposed on membranes become greater. Suspended precast floors offer little or no integral resistance to gas ingress due to joints etc; requiring greater emphasis on gas barrier
Installation of gas protection	All gas resistant membranes have sufficient gas impermeability. The main issue is surviving construction intact. Increased confidence in the installation of membranes is provided by the use of skilled installers, installation under CQA procedures, post-construction independent inspection of installations. "Note the minimum gauge unreinforced polyethylene membrane that is likely to survive construction intact is 2000 gauge"
	The effectiveness and robustness of the underfloor venting or proprietary pressurisation systems can be demonstrated by post construction independent validation of underfloor gas concentrations for all types of systems (including passive systems) and air flow rates (or positive pressures for proprietary systems). Continuous monitoring via a number of detection points is more reliable than a short period of intermittent sampling visits from a limited number of points.
Effectiveness of underfloor venting media	Air flow in all systems must flow through permeable media such as open voids, geocomposites or polystyrene void formers and gravel. The air flow characteristics determine the efficiency of the system and it is critical for all types of systems. Independent test data should be obtained from manufacturers to allow design calculations to be carried out for all types of venting and pressurisation media, using well-established methods. (Specific guidance on flow through venting media is provided in the Partners in Technology report (1997).
Effectiveness of membrane barriers	Different specific gas membranes have advantages and disadvantages. The main issue is surviving construction (see above).
Maintenance requirements	This refers to the extent to which gas protective measures rely on maintenance. Systems that require little maintenance are more robust. Simple monitoring regimes are best. Most active systems are more reliant on maintenance, but systems that can advise of fan and detection failure by alarm are more robust. Reliability of components affects risk and manufacturers can provide advice on run time life of fans. The risk of future component obsolescence should also be considered.

A6 Modelling and risk assessment for vapours

A6.1 INTRODUCTION

For the indoor inhalation pathway, most numerical models start from the known concentrations of volatile contaminants in soil or groundwater, as measured by laboratory analysis, predicting partitioning into the soil gas phase, transport through unsaturated soil (possibly including attenuation), entry into a building through its sub-structure and mixing with air in the ventilated interior space. Rather than being the endpoint, the estimated contaminant concentration at the point of exposure is usually used to evaluate risk by combination with human exposure frequencies and contaminant toxicity characteristics. Alternatively, tolerable intakes or index doses are used to back-calculate the concentrations of contaminants in soil or groundwater that would be protective of human health. Vapour intrusion models are highly complex, relying on a very large number of variables. Most of the parameters are not measured as a matter of routine in site characterisation investigations, and many are difficult to establish even when included as objectives in detailed risk evaluation investigations. Verification of model outputs is constrained by a general lack of comprehensive field data in the literature with which to calibrate the software. Sensitivity analyses are not always straightforward to perform and can be complicated by hierarchies of model input parameters and the possible ranges of realistic values. Models are also vulnerable to misuse. As a result, some concerns have been expressed about the validity of risk assessments that are reliant on vapour intrusion model outputs, and their use to support remedial action decisions. Several studies comparing the performance of various widely-used models have demonstrated variability in outputs spanning several orders of magnitude.

The indoor air pathway is predicted in different ways by the main assessment models in use in the UK, which are summarised in Box A6.1.

Box A6.1   *Examples of various methods in use for estimating vapour intrusion*

> **Generic risk assessment tools**
>
> The CLEA-2002 software, a probabilistic model that incorporates the Johnson and Ettinger model (1991), was used by the EA to calculate the soil guideline values (SGVs) for a set of standard land uses. SGVs can be adjusted for soil type. This software is to be replaced imminently by CLEA-UK, which will have "open architecture" to allow risk assessors to calculate site-specific assessment criteria for particular receptors and exposure patterns. The vapour intrusion algorithms have been replaced by the Johnson and Ettinger model (1991) in line with the findings of Evans *et al* (2002) and the guidance in CLEA Briefing Note 2 (Environment Agency, 2004c).
>
> The SNIFFER Tier 1 deterministic method (Ferguson *et al*, 2003), adopts a procedure based on the US risk based corrective action (RBCA) framework (ASTM, 2002), but is compatible with the UK CLEA framework, population characteristics and standard land uses.
>
> **Site-specific risk assessment packages**
>
> The BP Risk Integrated Software for Clean-ups (BP-RISC) and the GSI Toolkit, were developed for use within the RBCA framework and include extrapolated versions of the Johnson and Ettinger model (1991).

The CLEA framework was introduced in 2002 (Defra and Environment Agency, 2002a, b, c, d) to provide methods for generic assessment of human health risks from contamination of soil that are consistent with the Part IIA legislation and accompanying guidance. Following this, CLR11 *Model Procedures for the management of land contamination* (Defra and Environment Agency, 2004a) was also released which provides additional guidance on the structure of risk assessments, appropriate cut-off points and level of detail required to demonstrate whether risks are acceptable or remediation is needed.

As part of the CLEA programme, Defra and the Environment Agency are publishing generic soil guideline values (SGV) for soil that would be protective of human health. These are intended to be trigger values for first tier screening. For example, if the estimated mean concentration of a contaminant of concern (taken as the 95th percentile upper confidence limit, US95) within a particular averaging area is below its SGV for that land use, the soil can be considered to be uncontaminated. If the concentration exceeds the SGV, further work is required, such as gathering more data, remediating the site/area or undertaking a site-specific detailed quantitative risk assessment (DQRA).

In the absence of published SGVs, risk assessors and regulators are expected to calculate their own assessment criteria using published toxicological profiles or the protocol for assessing toxicity (Defra and Environment Agency, 2002c). However, this has created some uncertainty or confusion. The highest level guidance on undertaking risk assessments (DETR, 2000b) is all that is authoritative. While relevant, it leaves many in the contaminated land sector trying to find cost-efficient ways of deriving their own screening values from first principles. Others often turn to inappropriate foreign guideline values. Practical guidance has been issued by the Environmental Industries Commission jointly with the Association of Geotechnical and Geoenvironmental Specialists (EIC and AGS, 2004) to simplify screening assessments. Its use requires some caution, particularly where comparisons are made with guidance from outside the UK.

SGVs have been published for two organic contaminants only, toluene and ethylbenzene (Defra and Environment Agency, 2004b and 2005a). Delays have resulted from a programme of additional intermediate research into organic contaminants (Earl, 2003; Earl *et al*, 2003) to address uncertainties in specific pathways, for example plant uptake and, of particular interest, vapour intrusion (Environment Agency, 2004c).

The SNIFFER Tier 1 workbook (Ferguson *et al*, 2003), allows a separate calculation of the dilution ratio between concentrations of contaminants in soil under a building and in the indoor air, but defaults to a simplistic approach as adopted by the Swedish Environmental Protection Agency, using a uniform value of 1/20 000.

Guideline values are for soil only. If contamination is present in groundwater or exists as free-phase non-aqueous phase liquid (NAPL), its potential effects need to be assessed from first principles in a DQRA.

DQRAs often rely on the use of readily-available proprietary software models that calculate integrated indoor air pathways. The vapour intrusion element of most are based on extrapolations of the US Johnson and Ettinger Model (1991).

Golder Associates (Evans *et al*, 2002) reviewed and tested several software models on behalf of the Environment Agency during the later stages of development of the CLEA package. The study concluded that the version of the Johnson and Ettinger Model was the most suitable. In 2004, the Environment Agency introduced new guidelines on how to evaluate the indoor inhalation pathway in assessments of risks to human health from soil contamination, replacing a significant section of the 2002 CLEA guidance (certain sections of Chapter 6 and the default soil properties in Table 5.2). This briefing note also sets out the move to the Johnson and Ettinger approach.

The latest update of the CLEA software has replaced earlier models with the Johnson and Ettinger model and incorporate "open architecture" to allow risk assessors to change model parameters for DQRAs (Earl, 2003). No information is yet available on what parameters might be used as defaults to derive generic criteria, but research into building parameters by the Building Research Establishment (BRE) has resulted in CLEA Briefing Note 3 *Update of supporting values and assumptions describing UK building stock* (Environment Agency, 2004). The problem with the Johnson and Ettinger model is that it models an *in-situ* concrete floor (ie without an underfloor void). In the UK suspended floors with an underfloor void are very common (more than 50 per cent of housing is constructed with this type of ground floor) and this will have a significant impact on the risk of vapour intrusion.

A6.2 RISK EVALUATION MODELS FOR VAPOURS

"Measure if you can, model if you must", is the maxim often recited by engineers and scientists. So why model vapour intrusion?

Human health risk assessments are carried out to determine if people are being harmed or if there is a possibility that they might become harmed. The assessments are either generic, applying guideline values to widely applicable, standardised, simplified, conservative models of exposure, or site-specific, in which assessment criteria are derived based on the characteristics of particular contaminant properties, soils and building characteristics.

In generic screening assessments, actual indoor air concentrations, to be robust and reliable, would require far more effort to obtain than demanded by the purpose and accuracy of a screening exercise. In site-specific risk assessments, reliable indoor air concentrations would be ideal. The problem is that the variability of concentrations over time (diurnal, seasonal and unpredictable) dictates the need to monitor over lengthy periods to generate values that have a high attached degree of confidence. However, the fact that a site-specific risk assessment is being undertaken suggests that there might be a real problem under investigation, in which case, the investigation period may lead to extending the occupants' exposure and increasing the risk of harm. Lengthy monitoring periods can also mean increased costs in investigation, the value of which is not always appreciated.

There are also technical difficulties. Dawson (2002) outlined the dilemma that soil and groundwater sampling and analysis is most reliable but the sampling points are furthest from the receptors. Bag sampling bulk gas at the point of exposure (or even from soil gas) is less reliable for characterisation because of the effects of non-soil sources, pressure fluctuations, attenuation, and sampling error due to the equipment required. Analysis of gas is also more difficult, especially since contaminants will be at low (even if harmful) concentrations.

In Figure A6.1, the monitoring results for one year at a US military site indicates the potential magnitude of variability but also indicates the effect of spatial variations (Dawson, 2002).

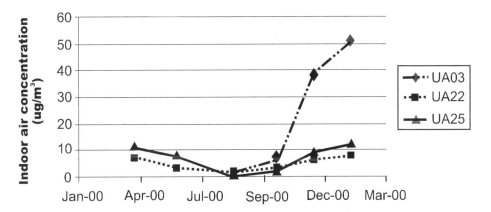

| Figure A6.1 | *Temporal and spatial variability in vapour concentrations* |

A6.2.1 Monitoring

The US Interstate Technology and Regulatory Council (ITRC), is a state-led coalition of regulators who work with industry and stakeholders to further technical knowledge in new fields and streamline regulatory acceptance of environmental technologies, which has researched the vapour intrusion problem. It found that indoor air quality was overlooked altogether in many investigations, and in those that recognised the indoor air pathway, lack of knowledge about monitoring techniques and considerations of cost, contributed to the wilful omission of indoor air sampling and evaluation (ITRC, 2004).

Although air sampling and analysis techniques have not always been adequate to detect the low concentrations of some contaminants in air that present risks to human health, techniques have advanced (ITRC, 2004). Tindle (2002) presented extensive revisions to a 1986 guidance document on analysing gas samples for the assessment of occupational exposure to gasoline to take account of recent developments. Hers (2000) conducted experiments specifically targeted at the measurement of vapour intrusion from ground contamination. His team constructed a new temporary building for which all parameters were recorded at a hydrocarbon contaminated site that had been well characterised. Monitoring was carried out in the new building and at an existing building on the site to test and advance measurement techniques.

Zampolli *et al* (2004) have developed a proposal for an "electronic nose", which employs metal oxide sensors and a fuzzy logic system to recognise patterns in indoor air quality associated with mixtures of gases and relative humidity. In an operating trial, they established success in discriminating low concentrations of target compounds. Riley *et al* (2003) have developed a small portable biological device using cultured cells from lungs as the recognition element for direct toxicity tests. They claim that its performance compares favourably with more traditional laboratory cell culture measurements.

Johnson *et al* (2002) made good use of field data after the incident at Colorado, calculating actual attenuation ratios (α) from indoor air and soil gas measurements, which they then compared to model outputs as a verification exercise.

A6.2.2 Modelling objectives

Numerical models provide an analytical solution to the simplified conceptual model representing what is known or can be assumed about the true situation. As such, the main objective is to obtain a sufficient set of output data to enable decision makers to make informed choices.

When conducted properly, modelling should provide a framework for discussion and should not in itself preclude further assessment options (ITRC, 2003). Although numerical models can make accurate predictions, they are only ever as good as the representation made by the conceptual model and the data used as input. Great care should be exercised to ensure that they are used appropriately.

A6.2.3 Model availability and selection

Recent reviews made in the literature

Some recent key reports that were commissioned by regulators, industry, standards organisations and technical issue networks, describe studies that were undertaken with the general aim of evaluating and comparing models of human exposure to land contamination, either for benchmarking and improving performance or for selecting an appropriate model to generate screening guideline values (Kennedy *et al*, 1998; Evans *et al*, 2002; Poletti *et al*, in press).

It was not an objective of this publication to question the findings of these studies or to re-appraise the source literature referenced in each. Where the authors have short-listed or selected models, it has been taken as read that their evaluation criteria were appropriate and their conclusions valid. It is the intention, however, to gain a brief insight into the methods of evaluation adopted and to establish the features of the models tested that were significant in comparisons with each other, with field data and in sensitivity analyses.

Poletti *et al* (in press)

The industrial sub-group of the network of industrially contaminated land in Europe (NICOLE) commissioned Arcadis Gerraghty and Miller to critically appraise the human health risk assessment systems and models commonly used in Europe. The purpose was to raise awareness and improve understanding of model variability with risk assessors, and to provide confidence to risk managers who rely on model outputs for business decisions. The key finding of Poletti *et al* that, of all exposure pathways, indoor air inhalation showed the greatest variation, with the doses predicted from the same data sets spanning three orders of magnitude was of particular interest.

The Arcadis team conducted a literature review and identified risk assessment methods being used in different European countries, screened the system capabilities of models, and then made comparative studies using test data. A brief summary is given here, with a particular focus on the vapour inhalation modelling.

The review sought to find previous studies with similar objectives and then to assess the evaluation methods and conclusions drawn. Six previous studies were cited as being relevant:

1   *Ferguson and Kasamas (eds) (1999)* – the two volume text commissioned by the network for Concerted Action on Risk Assessment for Contaminated Sites in Europe (CARACAS) to describe risk assessment practices in 16 European countries.

Particular reference was made to the matrices in the Volume 2 appendices that compared the various risk assessment methods.

2 *Zaleski and Gephart (2000)* – the Exposure Factors sourcebook for Europe, developed on behalf of NICOLE. Although not a comparison between different risk assessment methods or models, the pertinent statistics for the populations of different countries demonstrate the potential variability in generic parameters and hence demonstrate the need to use region-specific data.

3 *Whittaker et al (2001)* – a UK Environment Agency publication describing a method for benchmarking software models that predict water quality impacts from contamination. The comparison method was developed to accompany the Agency's method for deriving remedial targets for soil and groundwater to protect water resources (Marsland and Carey, 1999), and used case studies to assess model performance and generate outputs for comparison. Although not of direct relevance to vapour intrusion modelling, this study developed a useful robust framework for appraising varying approaches to contaminant transport and the modelling of flux between environmental media.

4 *Rikken et al (2001)* – part of the output of a Dutch research programme by RIVM to benchmark the country's CSOIL model and identify improvements. Focus was placed on three exposure pathways, one of which was indoor inhalation of volatile contaminants. A more detailed review is made of this report below.

5 *Evans et al (2002)* – another UK Environment Agency publication. This dealt specifically with vapour intrusion models, with the aim of identifying the most appropriate available model to incorporate into the CLEA system. A more detailed review is made of this report below.

6 *Swartjes (2002)* – a presentation made to disseminate the findings of another RIVM study into a wider comparison of risk assessment methods and models, particularly the variations arising from default parameters and hard-wired steps in algorithms. A review of an earlier published output of the same work (Swartjes, 2001) is given below.

The comparison exercise was phased. Firstly, a generic data set for a hypothetical site was run with the same input parameters, where possible, followed by the testing of sensitivity by varying certain input parameters in turn. Secondly, real data sets from selected sites were used to test the range of predictions from the models where default values were used for chemical and toxicological parameters, and for other exposure data where site-specific values could not be entered.

Evans *et al* (2002)

The Environment Agency engaged Golder Associates to provide guidance on the suitability of vapour intrusion models and to recommend an appropriate model to be used in the CLEA package. Evans *et al* reviewed 10 models, short-listing five for detailed analysis: the GSI RBCA toolkit, the British Columbia model, Volasoil (RIVM), BP RISC and the FKM/KF model.

The outputs from the five models were calibrated with site-specific data, using benzene as the indicator contaminant. The authors noted that no single model satisfied all the criteria expected to describe soil vapour intrusion but that the BP RISC version of the Johnson and Ettinger model came closest.

Rikken *et al* (2001)

Reporting to the Ministry of Housing, Spatial Planning and the Environment, the National Institute of Public Health and the Environment (RIVM) in The Netherlands reviewed the methods for deriving Dutch soil and groundwater intervention values with the model CSOIL after its first 10 years of use. The purpose was to appraise the methodology of CSOIL in light of up-to-date views on human exposure assessment and to evaluate its performance by comparison with other models. For comparison, Rikken *et al* used the European Union System for the Evaluation of Substances (EUSES), a tool designed to assess industrial emissions and which includes some indirect human exposure pathways, and three foreign models that deal specifically with human exposure to land contamination:

1   An early draft version of CLEA from the UK.
2   The German UMS.
3   CalTOX from the California State Department of Toxic Substances Control, USA.

Research was focused on the three exposure pathways that make the greatest contribution to the derivation of Intervention Values:

1   soil ingestion
2   consumption of contaminated crops
3   indoor inhalation of vapours.

The re-evaluation led to Rikken *et al* proposing modifications to improve each of the CSOIL assessment methods for these pathways. Concurrently, Otte *et al* (2001) reviewed model input parameters. Lizjen *et al* (2001) produced an integrated report with revised proposals for adjustments to CSOIL.

Kennedy *et al* (1998)

On behalf of the American Society for Testing and Materials (ASTM), Kennedy *et al* (1998) compiled a compendium of widely available non-proprietary fate and transport models that support risk-based corrective action (RBCA) by predicting contaminant movement and behaviour in the environment. The guidance note provided details of 39 models, covering pathways in all environmental media. No proprietary models or complex multi-media integrated models were included. Of the 39, three cover the soil to indoor air pathway and two the groundwater to indoor air pathway.

Summary of the findings

Where the studies resulted in firm conclusions or recommendations, the Johnson and Ettinger Model performed best. Of most relevance to the CLEA regime is the recommendation made by Evans *et al* (2002) that the Johnson and Ettinger Model sub-routine in BP RISC be taken as the model of choice for the Environment Agency.

It is unclear why the Agency did not act upon this recommendation sooner. The CLEA-2002 software emerged after the Golders report, with the FKM/KF models encoded. Only in 2004, has the Johnson and Ettinger Model been adopted (Environment Agency, 2004c), replacing sections of CLR10 (Environment Agency, 2002d).

A6.3 THE JOHNSON AND ETTINGER MODEL

A6.3.1 Background

The work of Johnson and Ettinger (1991) has been used as the basis for many subsequent vapour intrusion model packages.

The USEPA vapour intrusion approach (USEPA, 1997/2003, 2001c, 2002) is often referred to as an example of the application of the Johnson and Ettinger Model (1991). In fact, it is much more. The guidance adopts multiple lines of evidence approach to assessing the indoor inhalation pathway which, among other steps, includes a survey of household products, and sampling of indoor and outdoor air. The predictions using the modified JE model are one part of determining if the pathway is complete.

BP-RISC uses the soil gas to indoor air model in its groundwater to indoor air and soil to indoor air pathways (Spence and Walden, 2001). The Environment Agency (2004c) version is adapted from the BP RISC soil model.

A6.3.2 Details of the model

Johnson and Ettinger (1991) used a mass-transfer model that includes diffusion and advection to calculate the ratio (α) of steady-state contaminant concentration in indoor air to observed contaminant concentration in soil gas. According to their model, this alpha ratio is given by the following expression:

$$\alpha = \frac{\left[\dfrac{D_T^{eff} A_B}{Q_{building} L_T}\right] \times \exp\left(\dfrac{Q_{soil} L_{crack}}{D^{crack} A_{crack}}\right)}{\left[\exp\left(\dfrac{Q_{soil} L_{crack}}{D_{crack} A_{crack}}\right)\right] + \dfrac{D_T^{eff} A_B}{Q_{building} L_T} + \left(\dfrac{D_T^{eff} A_B}{Q_{soil} L_T}\right) \times \left[\exp\left(\dfrac{Q_{soil} L_{crack}}{D^{crack} A_{crack}}\right) - 1\right]}$$

in which:

D_T^{eff}	=	overall effective diffusion coefficient based on vapour phase concentration for the region between the source and the foundation (cm²/s)
A_B	=	cross-sectional area of the building foundation (cm²)
$Q_{building}$	=	building ventilation rate (cm³/s)
L_T	=	distance from the contaminant source to the foundation (cm)
Q_{soil}	=	volumetric flow rate of soil gas into the building (cm³/s)
L_{crack}	=	thickness of the building foundation (cm)
D_{crack}	=	effective vapour pressure diffusion coefficient through cracks in the foundation (cm²/s)
A_{crack}	=	the area of cracks through which contaminant vapours enter the building (cm²)

The conceptual model is shown in shown in Figure A6.2.

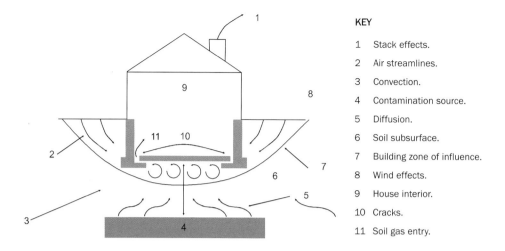

Figure A6.2 *Conception model based on the Johnson and Ettinger assumptions (USEPA, 1997/2003, and Environment Agency, 2004c)*

Many of the parameters in the primary algorithm are themselves functions of several secondary factors that need to be calculated or estimated (Johnson, 2002 and Schuver, 2003).

Fischer *et al* (2001) were engaged to peer review the approach developed by the Michigan Department of Environmental Quality for setting generic clean-up criteria. After considering many detailed factors in the methodology, they reached the conclusion that Johnson and Ettinger (1991) was appropriate for the indoor air inhalation pathway, being based in sound science, having wide acceptance by regulators and being cited by ASTM in the RBCA standard. Fischer *et al* (2001) claimed that the JE model was straightforward to use and that its simplicity and accompanying documentation allow easy site-specific modification. However, they did recognise that the complex problem being modelled requires an unusually large number of input parameters and that the range of variance of inputs could generate widely varying outputs. This appears to conflict with the understanding of the hierarchy of inputs and the reasonable ranges described by Johnson (2002), who claims that overall, variance should be small, with sensitivity displaying a linear response to changes in the primary parameters.

In a somewhat (understandably) defensive paper, Johnson *et al* (2002) describe the approaches adopted by some state departments, either relying on screening models only or insisting on indoor air monitoring, as being at the extremes of a spectrum of assessment methods. They give further support to Johnson's own recommendation for an intermediate approach combining a screening model with sensible range limits for some estimated parameters and some site-specific data for certain other parameters. They admit that the strengths and weaknesses of the model are difficult to assess objectively with such a scarcity of field data for validation. However, taking advantage of the large data set made available from the troublesome Colorado site, they go on to show how their analysis using reasonable estimates of primary input parameters would have compared well with the actual attenuation factors calculated from the data set.

A7 Situation B derivations of ground gas emissions to calculate gas screening values (GSVs) for low-rise housing development with a ventilated underfloor void (min 150 mm)

A7.1 MODEL LOW RISE HOUSING DEVELOPMENT

The NHBC (Boyle and Witherington, 2007) model for low-rise housing development is based upon a model low-rise house as shown in Figure A7.1. The property has a floor plan of 8.00 m × 8.00 m, giving a floor area of 64 m². Continuing the assumption that an individual 50 mm internal diameter borehole relates to an approximate area of 10 m², which was proposed by Pecksen (1986) and was further expanded by DETR (1997). This represents 6.4 boreholes for the floor plan of the house.

The model low-rise house was given the minimum recommended sub-floor void height of 0.15 m as specified in The Building Regulations "Approved Document C" (Department of the Environment and the Welsh Office, 2004 edition), which produces a sub-floor void space of 9.60 m³. As a worst-case scenario, it was considered that the ventilation rate within the sub-floor void was subject to a complete volume change every 24 hours, which is considered to be highly conservative. So the ventilation rate is 9.60/24 = 0.40 m³/hr.

For methane, the equilibrium concentration of gas within the sub-floor void is important as a stray ignition source (eg dropped cigarette or match) could ignite any accumulated methane, which could seriously affect life and property.

For carbon dioxide, however, the issue of the sub-floor is not as significant as humans will generally not enter this area of a property. What is of concern is if the membrane and/or floor above the sub-floor void are accidentally or otherwise penetrated, which would allow for a release of carbon dioxide into the house. For carbon dioxide a leak of gas from the sub-floor void into a small room (eg downstairs toilet with soil pipe potentially passing into sub-floor void) of dimensions 1.50 m × 1.50 m × 2.50 m with a total room volume of 5.63 m³ was considered. Again, as a worst-case scenario, it was considered that the ventilation rate of the room was subject to a complete volume change every 24 hours, which is again considered to be highly conservative. The ventilation rate within the room is 5.63/24 = 0.24 m³/hr. The leak from the sub-floor void was assumed to account for 10 per cent (0.024 m³/hr) of the small room ventilation rate, which is considered to represent a significant leak. This is considered to be highly conservative in its approach.

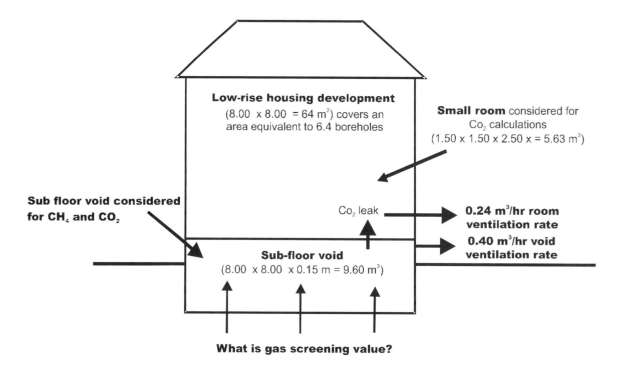

Figure A7.1 *Model low-rise residential property developed for calculating gas screening values for methane and carbon dioxide*

A7.2 METHANE GAS SCREENING VALUE DERIVATIONS

A7.2.1 Introduction

When the concentration of methane in air is between the limits of 5.0 % v/v and 15.0 % v/v an explosive mixture is formed with normal concentrations of oxygen. The lower explosive limit (LEL) of methane is 5.0 % v/v, which is equivalent to 100 per cent LEL. The 15.0 % v/v limit is known as the upper explosive limit (UEL), but concentrations above this level cannot be assumed to represent safe concentrations.

The maximum concentration of methane that could possibly be permitted in any sub-floor void is 100 per cent LEL (5.0 % v/v). However, these concentrations were considered by the NHBC (Boyle and Witherington, 2006) to represent an undesirable maximum and a more conservative maximum permissible equilibrium concentration of 2.5 % v/v methane (50 per cent LEL) within the sub-floor void was proposed. This still allows for worse ventilation rates to rise from those assumed on extremely still days. A concentration of 2.5 % v/v methane was considered to represent the upper threshold limit for development (red traffic light) and equilibrium concentrations above this would indicate that a standard low-rise housing development with passive ventilation systems would not be acceptable. A lower limiting equilibrium concentration of 1.0 % v/v methane within the sub-floor void was taken for the transition between amber 1 to amber 2.

The GSVs for methane are derived below.

A7.2.2 Amber 2 to red value

At equilibrium, the maximum concentration of methane considered allowable within the sub-floor void with protection prescribed in amber 2 to red is 2.5 % v/v. The maximum equilibrium rate of methane entering the sub-floor void is the same as the methane exiting the void, which is 2.5 per cent of 0.40 m³/hr = 0.01 m³/hr = 10 l/hr.

Assuming that the house occupies an area equivalent to 6.4 boreholes, the GSV for amber 2 to red traffic lights is:

10/6.4 = 1.56 l/hr.

The typical maximum concentration for amber 2 to red traffic lights is 20.0 % v/v.

If the GSV of methane is exceeded it would cause a Red Traffic Light and development should not continue, unless the gassing source can be removed or reduced, or agreement can be reached with the NHBC regarding other detailed protection measures including full legal agreements to cover maintenance and any other issues.

The typical maximum concentration may be exceeded if the GSV indicates it is safe to do so or a site specific GSV can be derived.

A7.2.3 Amber 1 to amber 2 gas screening value

At equilibrium, the maximum concentration of methane considered allowable within the sub-floor void with protection prescribed in amber 1 to amber 2 is 1.0 % v/v. The maximum equilibrium rate of methane entering the sub-floor void is the same as the methane exiting the void, which is 1.0 per cent of 0.40 m³/hr = 0.004 m³/hr = 4 l/hr.

Assuming that the house occupies an area equivalent to 6.4 boreholes, the GSV for amber 1 to amber 2 traffic lights is:

4/6.4 = 0.63 l/hr.

The typical maximum concentration for amber 1 to amber 2 traffic lights is 5.0% v/v.

If the GSV of methane is above this value it would cause an amber 2 traffic light and development should include protection measures as prescribed in amber 2. The typical maximum concentration may be exceeded if the GSV indicates it is safe to do so or a site specific GSV can be derived.

A7.2.4 Green/amber 1 gas screening value

If a sub-floor void were to be present, the maximum equilibrium concentrations of methane considered allowable would be 0.25 % v/v. The maximum equilibrium rate of methane entering a hypothetical sub-floor void would be the same as the methane exiting the void, which is 0.25 per cent of 0.40 m³/hr = 0.001 m³/hr = 1 l/hr.

Assuming that the house occupies an area equivalent to 6.4 boreholes, the GSV for green to amber 1 traffic lights is:

1/6.4 = 0.16 l/hr.

The typical maximum concentration for green to amber 1 traffic lights is 1.0 % v/v.

If the GSV is not exceeded no gas protection measures are considered necessary for methane. However, if the GSV of methane is above this value it would cause an amber 1 traffic light and development should include protection measures as prescribed in amber 1. The typical maximum concentration may be exceeded if the GSV indicates it is safe to do so or a site specific GSV can be derived.

A7.3 CARBON DIOXIDE GAS SCREENING VALUE DERIVATIONS

A7.3.1 Introduction

The UK Health & Safety Executive (HSE) has published information (HSE, 2002) relating to concentrations of carbon dioxide that humans may be exposed to, which uses concentrations contained in the Control of Substances Hazardous to Health (COSHH) Regulations 1999. These are the long-term exposure limit (eight hour period) and the short-term exposure limit (15 minute period), which are 0.5 % v/v and 1.5 % v/v carbon dioxide, respectively.

If an unknown penetration of the membrane above the sub-floor void occurred, continued release of carbon dioxide into the small room may happen. If this were to occur, the maximum permissible concentration of carbon dioxide within this room is considered to be at equilibrium 0.5 % v/v. It has been assumed that the leak from the sub-floor void will account for 10 per cent of the air in the small room, which is considered to represent a significant leak. This is considered to be highly conservative in its approach.

The GSVs for carbon dioxide are derived below.

A7.3.2 Amber 2 to red gas screening value

At equilibrium, the maximum concentration of carbon dioxide considered allowable within the sub-floor void with protection prescribed in amber 2 to red is not known. However, it is known that the maximum equilibrium concentration of carbon dioxide permissible within the small room is 0.5 % v/v.

As previously stated, it has been assumed that the ventilation of the small room is $5.63/24 = 0.24$ m^3/hr and that 10 per cent (0.024 m^3/hr) of this comes from the leak in the sub-floor void. At equilibrium, the carbon dioxide entering the small room is the same as the carbon dioxide exiting the small room, which is 0.5 % of 0.24 m^3/hr = 0.0012 m^3/hr. So the maximum concentration of carbon dioxide entering the small room is 5.0 % v/v (0.0012/0.024 × 100 %).

If the maximum carbon dioxide concentration entering the small room is also 5.0 per cent, the equilibrium concentration of carbon dioxide within the sub-floor void should also be 5.0 per cent, which is also the concentration of carbon dioxide exiting the sub-floor void. 5.0 %v/v of 0.40 m^3/hr = 0.02 m^3/hr = 20 l/hr.

Assuming that the house occupies an area equivalent to 6.4 boreholes, the GSV for amber 2 to red traffic lights is:

20/6.4 = 3.13 l/hr.

The typical maximum concentration for amber 2 to red traffic lights is 30.0 % v/v.

If the GSV of carbon dioxide is exceeded it would cause a red traffic light and development should not continue, unless the gassing source could be removed or reduced, or agreement can be reached with the NHBC regarding other detailed protection measures with full legal agreements to cover maintenance and other issues have been addressed.

The typical maximum concentration can be exceeded if the GSV indicates it is safe to do so or a site specific GSV can be derived.

A7.3.3 Amber 1 to amber 2 gas screening value

For the amber 1 to amber 2 traffic light it is considered that the maximum tolerable concentration of carbon dioxide within the small room is 50 per cent of the amber 2/red concentration, which is 0.25 % v/v. From the amber 2/red calculations, the maximum concentration of carbon dioxide entering the small room is 2.5 % v/v. The carbon dioxide exiting the void is 2.5 % v/v of 0.40 m^3/hr = 0.01 m^3/hr = 10 l/hr.

Assuming that the house occupies an area equivalent to 6.4 boreholes, the GSV for amber 2 to red traffic lights is:

10/6.4 = 1.56 l/hr.

the typical maximum concentration for amber 1 to amber 2 traffic lights is 10.0 % v/v.

If the GSV of carbon dioxide is above this value it would cause an amber 2 traffic light and development should include protection measures as prescribed in amber 2. The typical maximum concentration may be exceeded if the GSV indicates it is safe to do so or a site specific GSV can be derived.

A7.3.4 Green to amber 1 gas screening value

If a sub-floor void were to be present, which were to leak into a small room, the maximum equilibrium concentrations of carbon dioxide considered allowable within the small room would be 25 % of the amber 2/red concentration, which is 0.125 % v/v. From the amber 2/red calculations, the maximum concentration of carbon dioxide entering the small room equates to 1.25 % v/v of 0.40 m^3/hr = 0.005 m^3/hr = 5 l/hr.

Assuming that the house occupies an area equivalent to 6.4 boreholes, the GSV for green to amber 1 traffic lights is:

5/6.4 = 0.78 l/hr.

The typical maximum concentration for green to amber 1 traffic lights is 5.0 % v/v.

If the GSV is not exceeded no gas protection measures are considered necessary for carbon dioxide. However, if the GSV of carbon dioxide is above this value it would cause an amber 1 traffic light and development should include protection measures as prescribed in amber 1. The typical maximum concentration may be exceeded if the GSV indicates it is safe to do so or a site specific GSV can be derived.

Bibliography

ANDERSEN, C E, KOOPMAN, M and DE MEIGER, R J (1996)
Identification of advective entry of soil-gas radon into a crawl space covered with sheets of polyethylene foil
No 9486, BIBINF Denmark, Riso National Laboratory, 125 pp

AGS and EIC (2004)
Screening tool for human health risk assessment
<www.ags.org.uk/publications/CLEA-ScreeningTool.pdf>

APPLETON, J D (1997)
Radon, methane, carbon dioxide, oil seeps and potentially harmful elements from natural sources and mining areas: relevance to planning and development in Great Britain
BGS Technical Report WP/95/4. British Geological Survey, Keyworth

ASTM (1995/2002)
Standard Guide for Risk-Based Corrective Action Applied at Petroleum Release Sites
Active Standard: ASTM E1739-95(2002), E50.04, Book of Standards Volume: 11.05
<www.astm.org/cgi-bin/SoftCart.exe/DATABASE.CART/REDLINE_PAGES/E1739.htm?E+mystore>

BALL, B C, GLASBEY, C A and ROBERTSON, E A G (1994)
"Measurement of soil gas diffusivity in situ"
European Journal of Soil Science, **45** (1), pp 3–13
<www.blackwell-synergy.com/doi/abs/10.1111/j.1365-2389.1994.tb00480.x>

BARRY, D L, SUMMERSGILL, I M and GREGORY, R G (2001)
Remedial engineering for closed landfill sites
C557, CIRIA, London

BANNON, M P and HOOKER, P J (1993)
Methane: Its occurrence and hazards in construction
R130, CIRIA, London

BRE (1999)
Radon: Guidance on protective measures for new dwellings
BRE Report 211, BRE Press, Berkshire (ISBN: 1-86081-328-3)

CARD, G B (1996)
Protecting development from methane
Report 149, CIRIA, London

CAIRNEY, T (1987)
Reclaiming contaminated land
Blackie & Son, Glasgow

Chartered Institution of Waste Management (1987)
Monitoring of landfill gas

Chartered Institution of Waste Management (1998)
Monitoring of landfill gas – update
IWM Landfill Gas Monitoring Working Group, for the Institution of Waste Management

COLE, K (1993)
"Building over abandoned shallow mines. Paper 1: Considerations of risk and reliability"
Ground Engineering, **26** (1), Jan/Feb

CRIPPS, A (1995)
Time-dependent modelling of soil gas movement: a literature review
BRE Press, Berkshire (ISBN: 978-1-86081-057-2)

CRIPPS, A (1998)
Modelling and measurement of soil gas flow
BRE Report 338, BRE Press, Berkshire (ISBN: 978-1-86081-200-2)

CRIPPS, A (1999)
"Solutions to a mixed boundary problem for soil gas flow"
Building and Environment, 34, **6**, November, pp 721–726

CROFT, B and EMBERTON, R (1989)
"Landfill gas and the Oxidation of methane in soil"
The Technical Aspects of Controlled Waste Management, Research Report No. CWM 049/89, Oxfordshire, UK, Department of the Environment, Wastes Technical Division

CROWHURST, D (1987)
Measurement of gas emissions from contaminated land
BRE Press, Watford (ISBN: 0-85125-246-X). Now out of print

CROWHURST, D and MANCHESTER, S J (1993)
The measurement of methane and other gases from the ground
Report 131, CIRIA, London (ISBN: 978-0-86017-372-4)

DAWSON, H E (2002)
Evaluating vapour intrusion from groundwater and soil to indoor air
EPA Brownfields Conference, November. Available from:
<www.epa.gov/correctiveaction/eis/vapor/f02052.pdf>

Defra (2000)
Circular 02/2000, Contaminated Land

Defra (2000)
Model Procedures for the management of land contamination
CLR11, Defra and Environment Agency (ISBN: 1-84432-295-5) <www.environment-agency.gov.uk/commondata/105385/model_procedures_881483.pdf>

Defra (2000)
Guidelines for environmental risk assessment and management
<www.defra.gov.uk/ENVIRONMENT/risk/eramguide/index.htm>

Defra (2002)
Waste management paper 27
Department of the Environment, November 2002

Defra (2004)
Review of environmental and health effects of waste management: municipal solid waste and similar wastes
<www.defra.gov.uk/environment/waste/research/health/pdf/health-report-contents.pdf>

Defra and EA (2002)
The contaminated land exposure assessment model (CLEA): Technical basis and algorithms
R&D Publication CLR10 Environment Agency, Swindon

DEPARTMENT OF THE ENVIRONMENT (1989)
Landfill sites: Development control
DoE Circular No. 38/89, 26 July

DOMENICO, P A and SCHWARTZ, F W (1997)
Physical chemical hydrogeology 2nd edition
John Wiley & Sons, New York, USA (ISBN: 978-0-471-59762-9)

DRAXLER, R R (1981)
Fifty-eight hour atmospheric dispersion forecasts at selected locations in the United States
NOAA Technical Memorandum ERL ARL-100

EDWARDS, S J and HUISH, N (1996)
"The study of hazardous subsurface gases by the use of automatic data logging equipment"
In: *Proc 4th int conf construction on polluted and marginal land, Brunel University, London, 2–4 July 1996*

ENVIRONMENT AGENCY (2001)
Guide to good practice for the development of conceptual models and the selection and application of mathematical models of contaminant transport processes in the sub surface
Report NC/99/38/2, National groundwater and contaminated land centre
<www.sepa.org.uk/pdf/cont_land/partii/good_practice_guide.pdf

ENVIRONMENT AGENCY (2003)
Impact assessment of landfill gas management guidance
EIA study report
<www.epd.gov.hk/eia/register/report/eiareport/eia_0892003/eia_report/Report_Section9.htm>

ENVIRONMENT AGENCY (2003)
Consultation on Agency Policy: Building Development on or within 250m of a landfill site
<www.environment-agency.gov.uk/commondata/acrobat/gdpo_con_aug_03_529983.pdf>

ENVIRONMENT AGENCY (2003)
Environment Agency guidance on monitoring surface emissions
Draft for consultation <www.environment-agency.gov.uk/commondata/acrobat/lftgn07_surface_936575.pdf>

ENVIRONMENT AGENCY (2003)
Guidance on monitoring leachate, groundwater and surface water
R&D project HOCO_232, Environment Agency
<www.sepa.org.uk/pdf/guidance/landfill_directive/monitoring_leachate_ground_surface_guidance.pdf>

ENVIRONMENT AGENCY (2003)
Consultation on Agency Policy: Building development on or within 250 metres of a landfill site – background information
<www.environment-agency.gov.uk/yourenv/consultations/529972/?lang=_e&theme=&region=&subject=&searchfor=outlines&any_all=&choose_order=&exactphrase=&withoutwords=>

ENVIRONMENT AGENCY (2003)
Principles for evaluating the human health risks from petroleum hydrocarbons in soils: A consultation paper
R&D Technical Report P5-080/TR1
<www.environment-agency.gov.uk/commondata/acrobat/petroleum_hydrocarbons1.pdf>

ENVIRONMENT AGENCY (2004)
Guidance on landfill completion. A consultation by the Environment Agency
<www.environment-agency.gov.uk/commondata/acrobat/landfill_guidance.pdf>

ENVIRONMENT AGENCY (2004)
Guidance on monitoring trace components of landfill gas
<www.sepa.org.uk/pdf/guidance/landfill_directive/trace_components_landfill_gas.pdf>

ENVIRONMENT AGENCY (2004)
Guidance on the assessment of risks from landfill sites
External Consultation, Version 1.0
<www.environment-agency.gov.uk/commondata/acrobat/risk_a_landfills__v1_768278.pdf>

ENVIRONMENT AGENCY (2004)
Update on estimating vapour intrusion into buildings
CLEA Briefing Note 2
<www.environment-agency.gov.uk/commondata/acrobat/bn2_904791.pdf>

ENVIRONMENT AGENCY (2004)
Update of supporting values and assumptions describing UK building stock
CLEA Briefing Note 3
<www.environment-agency.gov.uk/subjects/landquality/113813/672771/675330/678753/?version=1&lang=_e>

EREMITA, P (2000)
Edited/adapted field guideline for protecting residents from inhalation exposure to petroleum vapors
State of Maine Department of Environmental Protection

EVANS, D, *et al* (2002)
Vapour transport of soil contaminants
R & D Technical Report P5-018/TR, Environment Agency, (ISBN: 1-85705-151-3)

EXXON BIOMEDICAL SCIENCES INC *et al* (1997)
Human health risk-based evaluation of petroleum release sites: Vol. 4 - development of fraction specific reference doses and reference concentrations for total petroleum hydrocarbons
Total Petroleum Hydrocarbon Criteria Working Group, Association for the Environmental Health of Soils (AEHS) (ISBN: 1-884-940-13-7) <www.aehs.com>

FERGUSON, C C and DENNER, J M (1998)
"Human health risk assessment using UK Guideline values for contaminants in soil contaminated land and groundwater: Future directions"
Engineering Geology Special Publications, Geological Society, London, vol 14, pp 37–43

FERGUSON, C C, KRYLOV, V V and MCGRATH, P T (1995)
"Contamination of indoor air by toxic soil vapours: a screening risk assessment model"
Building and Environment, 30, **3**, July, pp 375–383

FISCHER, M L *et al* (1996)
"Factors affecting indoor air concentrations of volatile organic compounds at a site of subsurface gasoline contamination"
Environmental Science and Technology, 30, **10**, pp 2948–2957

FISCHER, L J *et al* (2001)
Evaluation of the Michigan Department of Environmental Quality's generic groundwater and soil volitization to indoor air inhalation criteria
MESB

FITZPATRICK, N A and FITGERALD, J J (1996)
"An evaluation of vapor intrusion into buildings through a study of field data"
In: *Proc 11th Annual Conference on Contaminated Soils, University of Massachusetts at Amherst, October 1996*. Go to: <www.mass.gov/dep/cleanup/gw2proj.pdf>

FOLKES, D J and ARELL, P S (2003)
"Vapor intrusion – EPA's new regulatory initiative and implications for industry"
In: *Proc Environmental Litigation Seminar, Snowmass, Colorado, UA, 18*

GODSON, J A E, WITHERINGTON, P J and McENTEE, J M (1996)
"Evaluation of risk associated with hazardous ground gases"
In: *Proc of the 4th International Conference, Construction on Polluted and Marginal Land, Brunel University, London, 2–4 July 1996*

GOWERS, A and COLEMAN, P (2004)
A review of the methodology for deriving ambient air level goals (AALGs) proposed by Calabrese and Kenyo
R&D Technical Report P6-020/1/TR1, Environment Agency

GREGORY, R G, *et al* (2000)
"A framework to assess the risks to human health and the environment from landfill gas (HELGA)"
Technical Report P271 (under contract CWM 168/98), Environment Agency

GUO, H, LEE, S C, CHAN, L Y and LI, W M (2004)
"Risk assessment of exposure to volatile organic compounds in different indoor environments"
Environmental Research, Vol 94, pp 57–66

GUSTAFSON, J *et al* (1997)
Human health risk-based evaluation of petroleum release sites: Vol. 3 - selection of representative tph fractions based on fate and transport considerations
Total Petroleum Hydrocarbon Criteria Working Group, Association for the Environmental Health of Soils (AEHS) (ISBN: 1-884-940-12-9) <www.aehs.com>

HARRIS, C R, WITHERINGTON, P J and McENTEE, J M (1995)
Interpreting measurements of gas in the ground
R151, CIRIA, London

HARTLESS, R (1991)
Construction of new buildings on gas-contaminated land
BRE Report 212, BRE Press, Berkshire (ISBN: 0-85125-513-2)

CONSULTANTS IN ENVIRONMENTAL SCIENCES LTD (1996)
Modelling site landfill gas resources: Best practice case studies
ETSU Report B/EW/00497/10/REP, Harwell Laboratory, 33 pp

HERS, I (2000)
"Measurement of BTX vapour intrusion into an experimental building"
In: *Proc presentation at the United States Environmental Protection Agency (USEPA) Resource Conservation and Recovery Act (RCRA) Corrective Action Environmental Indicator Forum*

HERS, I, ZAPF-GILJE, R, EVANS, D and LI, L (2002)
"Comparison, validation, and use of models for predicting indoor air quality from soil and groundwater contamination"
Soil and sediment contamination, 11, **4**, CRC Press

HERS, I, ZAPF-GILJE, R, EVANS, D and LI, L (2003)
"Evaluation of the Johnson and Ettinger Model for prediction of indoor air quality"
Ground Water Monitoring and Remediation, 23, **1**), pp 62–76

HERS, I, ZAPF-GILJE, R, LI, L and ATWATER, J (2001)
The use of indoor air measurements to evaluate intrusion of subsurface VOC vapors into buildings
Journal of the Air and Waste Management Association, **51**, pp 1318–1331

HOOKER, P J and BANNON, M P (1993)
Methane: its occurrence and hazards in construction
R131, CIRIA, London

HSE (2002)
Occupational exposure limits and published odour thresholds
Guidance Note EH40, Health & Safety Executive

Institute of Petroleum (1998)
"Guidelines for investigation and remediation of petroleum retail sites"
In: *Guidelines identifying the stages of a site investigation and appropriate remediation techniques*, Colchester, UK, Portland Press

THE INTERSTATE TECHNOLOGY & REGULATORY COUNCIL (ITRC) (2003)
Vapor intrusion issues at brownfield sites
Prepared by Brownfields Team <www.itrcweb.org/Documents/BRNFLD-1.pdf>

Interstate Technology & Regulatory Council (ITRC) (2004)
Brownfields Team Regulatory acceptance for new solutions: Vapour intrusion (indoor air)
Available from: <www.itrcweb.org> (accessed 2 July 2004)

JACOBS, J AND SCHARFF, H (1990)
Comparison of methane emission models and methane emission measurement
NV Afvalzorg, The Netherlands. Available from:
<www.fead.be/downloads/(Comparison_of_methane_emission_models_to_methane_emission__205).pdf>

JOHNSON, P C (2002)
Identification of critical parameters for the Johnson and Ettinger
Vapour intrusion model 1991, Bulletin 17, American Petroleum Institute, USEPA

JOHNSON, P C, KEMBLOWSKI, M W and JOHNSON, R L (1999)
"Assessing the significance of subsurface contaminant migration to enclosed spaces: site-specific alternatives to generic estimates"
Journal of Soil Contamination, 8, **3**, pp 389–421

JOHNSON, P C and ETTINGER, R A (1991)
"Heuristic model for predicting the intrusion rate of contaminant vapors into buildings"
Environmental Science and Technology, **25**, pp 1445–1452

JOHNSON, R (2001)
Protective measures for housing on gas-contaminated land
BRE Report 414, BRE Press, Berkshire (ISBN: 1-86081-460-3)

KENNEDY, J, MARINICIO, R, CARAVATI, M and KENNEDY, E (1998)
RBCA Fate and transport models: Compendium and selection guidance
Report by Foster Wheeler Environmental Corporation to the American Society for Testing and Materials

KESKIKURU, T, KOKOTTI, H, LAMMI, S and KALLIOKOSKI, P (2001)
"Effect of various factors on the rate of radon entry into two different types of houses",
Building and Environment, 36, **10**, December, pp 1091–1098

KRYLOV, V V and FERGUSON, C C (1998)
"Contamination of indoor air by toxic soil vapours: the effects of sub-floor ventilation and other protective measures"
Building and Environment, 33, **6**, pp 331–347

KURZ, D W (2000)
"Estimating residential indoor air impacts due to groundwater contamination"
In *Proc conf Hazardous waste research: Environmental challenges and solutions to resource development, production, and use, Denver, USA, May 23–25, 2000*

LITTLE, J C, DAISEY, J M and NAZAROFF, W W (1992)
"Transport of subsurface contaminants into buildings: an exposure pathway for volatile organics"
Environmental Science and Technology, **26**, pp 2058–2066

McMAHON, A, CAREY, M, HEATHCOTE, J and ERSKINE, A (2001)
Guidance on the assessment and interrogation of subsurface analytical contaminant fate and transport models
Report NC/99/38/1, Environment Agency National Groundwater and Contaminated Land Centre

McMAHON, A, CAREY, M, HEATHCOTE, J, ERSKINE, A and BARKER, J (2001)
Guidance on assigning values to uncertain parameters in subsurface contaminant fate and transport modelling
Report NC/99/38/3, Environment Agency National Groundwater and Contaminated Land Centre

MDEQ (1998)
Generic groundwater and soil volatilization to indoor air inhalation criteria: Technical support document
Michigan Department of Environmental Quality Environmental Response Division
<www.deq.state.mi.us/documents/deq-erd-tsd5.pdf>

NEY, R E (1990)
Where did that chemical go? A practical guide to chemical fate and transport in the environment
Van Nostrand Reinhold, New York (ISBN: 978-0-44200-457-6)

NHDES (1998b)
Residential indoor air assessmnet guidance document, draft
NHDES, Waste Management Division, Concord, NH

NHBC (2000)
Guidance for the safe development of housing on land affected by contamination
<publications.environment-agency.gov.uk/pdf/SR-DPUB66-e-e.pdf>

ODPM (2004)
Approved Document C: Site preparation and resistance to contaminants and moisture
The Stationary Office, Norwich (ISBN: 978-1-85946-202-7)
<www.planningportal.gov.uk/uploads/br/BR_PDFs_ADC_2004.pdf>

ODPM (2004b)
Planning Policy Statement 23: *Planning and pollution control*
The Stationary Office, Norwich
<www.planninghelp.org.uk/NR/rdonlyres/ 556E0840-1586-4B14-807F-141BD4F39A67/0/PPS23.pdf>

OLSON, D A and CORSI, R L (2001)
"Characterising exposure to chemicals from soil vapor intrusion using a two-compartment model"
Atmospheric Environment, **35** pp 4201–4209

O'ORIORDAN, N J and MILLOY, C J (1995)
Risk assessment for methane and other gases from the ground
R152, CIRIA, London

OTTE, P F, LIZJEN J P A, OTTE, J E, SWARTJES, F A and VERSLUIJS, C W (2001)
Evaluation and revision of the CSOIL parameter set. Proposed parameter set from human exposure modelling and deriving intervention values for the first series of compound
RIVM Rapport 711701021 <www.rivm.nl/bibliotheek/rapporten/711701021.html>

PARK, H S (1999)
"A method for assessing soil vapour intrusion from petroleum release sites: Multi-phase-fraction partitioning"
Global Nest: the Int. J. Vol **3**, pp 195–204

PARKER, J C (2003)
"Modeling volatile chemical transport, biodecay and emission to indoor air"
Ground Water Monitoring & Remediation, 23, **1**, pp 107–120

PARTNERS IN TECHNOLOGY (1997)
"Passive venting of soil beneath buildings"
Guide for Design, Vol 1 and 2, DETR, September

POLETTI, E *et al* (2003)
Risk assessment comparison study
Final Draft Report 916830024 to NICOLE/ISG (Network for Industrially Contaminated Land in Europe Industrial Sub-Group) (in press), Arcadis Geraghty & Miller International Incorporated, July

POLSON, C J and MARSHALL, T K (1974)
The disposal of the dead
3rd edn, Hodder Arnold (ISBN: 978-0-34016-247-7)

POTTER, T L and SIMMONS, K E (1998)
Human health risk-based evaluation of petroleum release sites: Vol. 2 - composition of petroleum mixtures
Total Petroleum Hydrocarbon Criteria Working Group, Association for the Environmental Health of Soils (AEHS) (ISBN: 1-884-940-19-6) <www.aehs.com>

RAYBOULD, J G, ROWAN, S P and BARRY, D L (1995)
Methane investigation strategies
R150, CIRIA, London

RAW, G J, COWARD, S K D, BROWN, V M and CRUMP, D R (2004)
"Exposure to air pollutants in English homes"
Journal of Exposure Analysis and Environmental Epidemiology, **14**, pp S85–S94

RIKKEN, M G J, LIZJEN, J P A and CORNLESSE, A A (2001)
Evaluation of model concepts on human exposure
RIVM Report 711701022, The Netherlands National Institute of Public Health and the Environment

ROBINSON, N (2003)
"Modelling the migration of VOCs from soils to dwelling interiors"
In: *Proceedings of the 5th national workshop on the assessment of site contamination*, A Langley, M Gilbey, and B Kennedy (eds), 47-712003
<www.ephc.gov.au/pdf/cs/workshopdocs/04_TPHs_Robinson_Migration_VOCs.pdf>

ROBINSON, A L and SEXTRO, R G (1997)
"Radon entry into buildings driven by atmospheric pressure fluctuations"
Environmental Science and Technology, 31, **6**, pp 1742–1748

RODRICKS, L A and CARON, J (2002)
"Methodology for evaluating vapor transport from free phase versus dissolved phase petroleum hydrocarbon (PHC) mixtures in the subsurface and its significance to indoor air quality"
In: *Proc meeting paper presented at Society for risk analysis, CH2M Hill Canada Ltd*
<www.riskworld.com/Abstract/2002/SRAam02/ab02aa240.htm>

RUDLAND, D J, LANCEFIELD, R M and MAYEL, P N (2001)
Contaminated land risk assessment – Guide to good practice
C552, CIRIA, London

RUSSELL, D L (1992)
Remediation Manual for petroleum-contaminated sites
Technomic Publishing Co Inc, Lancaster, USA

RYAN, G, KING, P J and MUNDAY, G (1988)
Report of the non-statutory public inquiry into the gas explosion at Loscoe, Derbyshire, 24 March 1986
Derbyshire County Council

RYDOCK, J and SKARET, E (2002)
"A case study of sub-slab depressurization for a building located over VOC-contaminated ground"
Building Environment, **37**, pp 1343–1347

SCHUVER, H J (2003)
"Overview of the science behind USEPA's guidance for the vapour intrusion to indoor air pathway"
In: *Proc USEPA Indoor Air Vapor Intrusion Seminar, 14 January*

SCOTT, P E, DENT, C G, and BALDWIN, G (1988b)
The composition and environmental impact of household waste derived landfill gas: second report
Report AERE G4436 Waste Research Unit, AERE Harwell, Environment Agency Report no CWM 041/88

SITE INVESTIGATION STEERING GROUP (1993)
Site investigation in construction Part 4: Guidelines for the safe investigation by drilling of landfills and contaminated land
Thomas Telford Ltd, London (ISBN: 978-0-72771-985-0)

SIZER, K, CREEDY, D P and SCEAL, J S (1996)
Methane and other gases from disused mines: The planning response
Technical report prepared by Wardell Armstrong for the Department of the Environment under Research Contract DoE PECD 7/1/445, HMSO, 189 pp

SLADEN, J A, PARKER, A and DORRELL, G L (2001)
"Quantifying risk due to ground gas on brownfield sites"
Land Contamination & Reclamation, 9, **2**, pp 191–208

STERRITT, R M (1995)
A review of landfill gas modelling techniques
In: Department of Trade and Industry (DTI), Energy Technology Support Unit (ETSU), UK, ETSU B B4 00413 REP, 82 pp

TINDLE, P E (2002)
Method for monitoring exposure to gasoline vapour in air – revision 2002
Report No 8/02, Concawe, Brussels

TRUESDALE, R S, LOWRY, M I and WOLF, S N (2002)
Draft procedure and issues report: Vapor intrusion pathway
RTI International Report to Indiana Department of Environmental Management
<www.spea.indiana.edu/msras/DraftVaporReport7-08-02.pdf>

TURCZYNOWICZ, L and ROBINSON, N (2001)
"A model to derive soil criteria for benzene migrating from soil to dwelling interior in homes with crawl spaces"
Human and Ecological Risk Assessment, 7, **2**, pp 387–415 (abstract only)

USEPA (1997/2003)
User's guide for the Johnson and Ettinger (1991) model for subsurface vapor intrusion into buildings
Contract No. 68-D30035, Prepared by Environmental Quality Management, Inc.,

USEPA (2001)
Correcting the Henry's Law constant for soil temperature
Factsheet. Available from:
<www.epa.gov/oswer/riskassessment/airmodel/pdf/factsheet.pdf>

USEPA (2001)
Henry's Law Constant
<www.epa.gov/athens/learn2model/part-two/onsite/esthenry.htm>, Athens

USEPA (2001)
RCRA Draft supplemental guidance for evaluating the vapor intrusion to indoor air pathway (vapor intrusion guidance)
<www.epa.gov/correctiveaction/eis/vapor/vapor.pdf>

USEPA (2002)
OSWER Draft guidance for evaluating the vapor intrusion to indoor air pathway from groundwater and soils (subsurface vapor intrusion guidance)
<www.epa.gov/correctiveaction/eis/vapor/complete.pdf>

VORHEES, D J *et al* (1999)
Human health risk-based evaluation of petroleum release sites: Vol. 5 - implementing the working group approach
Total Petroleum Hydrocarbon Criteria Working Group, Association for the Environmental Health of Soils (AEHS) (ISBN: 1-884-940-12-9) <www.aehs.com>

WANG, F and WARD, IC (1999)
"The development of a radon entry model for a house with a cellar"
Building and Environment, 35, **7(1)**, pp 615–631

WEISMAN, W (ed) (1998)
Human health risk-based evaluation of petroleum release sites: Vol. 1 - analysis of petroleum hydrocarbons in environmental media
Total Petroleum Hydrocarbon Criteria Working Group, Association for the Environmental Health of Soils (AEHS) (ISBN: 1-844-940-14-5) <www.aehs.com>

WHITTAKER, J J, BUSS, S R, HERBERT, A W and FERMOR, M (2001)
Benchmarking and guidance on the comparison of selected groundwater risk assessment models
Report NC/00/14, Environment Agency National Groundwater and Contaminated Land Centre

WILSON, S A and CARD, G B (1999)
"Reliability and risk in gas protection design"
Ground Engineering, 32, **2**, pp 32–36, February, EMAP, London

WISCONSIN DEPARTMENT OF HEALTH AND FAMILY SERVICES (2003)
Chemical vapor intrusion and residential indoor air: Guidance for environmental consultants and contractors
DHFS, Madison <dhfs.wisconsin.gov/eh/Air/fs/VI_prof.htm>

WITHERINGTON, P and BOYLE, R (2004)
Guidance on evaluation of development proposals on site where methane and carbon dioxide are present
Report no 4, National House Building Council and Risk Group Plc
<www.nhbcbuilder.co.uk/NHBCpublications/LiteratureLibrary/Technical/filedownload,29440,en.pdf>

WOOD, A A, GRIFFITHS, C M, BARRY, D L, RIDLEY A, O RIORDAN, N J and BECKETT, M J (1995)
"Contaminated sites are being over-engineered"
In: *Proc the Institution Of Civil Engineers,* Vol 108, 4, discussion of papers 10395/6, pp 80–185, November 1995

YOUNG, A (1990)
"Volumetric changes in landfill gas flux in response to variations in atmospheric pressure"
Waste Management and Research, **8**, pp 379–385

British Standards

BS 10381:2003 *Soil quality – sampling part 7 sampling of soil gas*

BS 5930:1999 *Code of practice for site investigations*

BS 10175:2001 *Investigation of potentially contaminated sites – Code of practice*